DEDICATED TO

DAVID B. STEINMAN
Designer of the Mackinac Bridge

PRENTISS M. BROWN
Chairman, Mackinac Bridge Authority 1950-1973

1887-1961

"The Bridge the Faith has built"

1889-1973

"The Father of the Mackinac Bridge"

Mighty Mac

The Official Picture History
of The Mackinac Bridge

by LAWRENCE A. RUBIN

with a Foreword by G. MENNEN WILLIAMS
GOVERNOR OF MICHIGAN

a Preface by PRENTISS M. BROWN
CHAIRMAN, MACKINAC BRIDGE AUTHORITY

an Introduction by FRANK B. WOODFORD

and an Essay on the Design of the Bridge
by DR. DAVID B. STEINMAN

Photographs supplied by HERMAN ELLIS
CHIEF PHOTOGRAPHER OF THE MACKINAC
BRIDGE AUTHORITY

 Wayne State University Press   Detroit

THE MACKINAC BRIDGE is a triumph of science and art.

It is a monument—an enduring monument—to vision, faith, and courage.

Without the vision, faith, and courage of the people of Michigan—their leaders, their statesmen, their workers—this great bridge could never have been built.

Outsoaring gravity and space, it rises from the waves on shining strands to arch across the sky in lofty grace.

This is our triumph over ancient fears.

A Bridge of Peace, wrought of the dreams of man!

D. B. STEINMAN

Foreword

THESE MAGNIFICENT PICTURES, which tell the story of the world's greatest bridge, make an historic record of one of mankind's most dramatic feats of engineering and construction.

This bridge across the Straits of Mackinac ranks with the pyramids, the great hydro-electric dams, the skyscrapers of Manhattan, the Panama and Suez canals, as one of the wondrous works of the hand and brain of man.

Yet no picture can fully tell the story of the Mackinac Bridge as a manifestation of the indomitable will of men to realize their dreams. This bridge stands across one of the world's great crossroads of commerce. Over this crossroads came the canoes of the explorers and the missionaries who brought civilization to the West. Through this area flowed the fur trade which was at once the motive and the means of opening the Great Northwest, and through these waters came the ships which carried the lumber to build America's cities.

For generations men dreamed of a land route across the Straits of Mackinac, a route which would constitute a new Northwest Passage, from the great cities of mid-America to the American and Canadian Northwest. Today, through the faith, the will, and the determination of the people of Michigan that dream is a reality.

The Mackinac Bridge is more than a gigantic physical structure, more than a mighty engineering feat, more than a link in the nation's new system of modern highways. The Mackinac Bridge is the manifestation, in steel and concrete, of the spirit of Michigan—a spirit adventurous, unafraid, respectful of the past but eager to meet the future, the spirit of a people for whom no task is too difficult, no job too big.

G. MENNEN WILLIAMS
Governor of Michigan

Preface

THE ENGINEERS and the contractors state that with proper maintenance the Mackinac Bridge will endure for at least a thousand years.

Right now, it is too soon to say whether or not a highway bridge, even of Mackinac's magnitude, will still be used as such ten centuries hence. In any event, its history should be recorded—and recorded in a manner that will impose the least difficulty upon future historians.

Actually, the history of the Mackinac Bridge can readily be divided into three parts. First, there is the intriguing narrative story that deals with the bridge background before the creation of the present Authority. Second are the experiences of the present Authority covering the period up to the bridge opening, unfolding in detail the trials and tribulations of the financing, legislation, contracting, engineering and construction.

The third part begins with November 1, 1957, the day the bridge was opened, and tells the story of the bridge operation—a story just starting, but one that promises to be as exciting and interesting as the prologue—all that happened before.

This volume, the second part of the Mackinac Bridge story, is devoted to the actual physical construction of the bridge. While this may well be told in words, it is fitting that it be told in pictures. And it is in pictures that the story of the four years of building the Mackinac Bridge is here presented.

The Authority, in carrying out its duties and responsibilities, did many things with originality and boldness not generally associated with a governmental agency. Fortunately, the Authority, made up of successful businessmen, was given the opportunity by the legislature to function with all the advantages of a state agency and without any of its restrictive disadvantages. Among the several distinctive bridge construction problems was that dealing with public relations, and this was solved by an agreement with three of the large contractors that they, along with the Authority, contribute to the expense of compiling a complete and detailed photographic history of the construction of the bridge. This photographic program operated under the direct control of the Authority but was financed in a large measure by the contractors. Through this cooperative procedure more than 3,000 black and white photographs were taken, about 1,000 color transparencies, and 14,000 feet of 16 mm commercial Kodachrome film for making documentary motion pictures of the construction. Two of these have already been produced and two others are in progress.

This volume is a collection of the most descriptive black and white photographs taken during the construction of the bridge. Each photograph, accompanied by a suitable caption, describes in detail and in layman's language a part of the complete story of how the bridge was built. It is the Authority's official pictorial history of the construction of the Mackinac Bridge. We trust it will be a companion volume to the official overall story of the Authority to be prepared in the near future and that it will someday be matched with the third part, telling the story of the bridge since its opening on November 1, 1957.

PRENTISS M. BROWN, *Chairman*
Mackinac Bridge Authority

Preface to the Third Printing

A whole new generation has grown up since the Bridge opening in November, 1957. It knows little of the legislative, financial and construction problems with which the Members of the Mackinac Bridge Authority were confronted before, during and after completion of the magnificent structure connecting Michigan's two beautiful peninsulas.

First, there were the financial problems that developed when traffic did not reach the estimates predicted and revenues were insufficient to meet the high levels required under the terms of the contract with the bondholders. They pressured the Authority to raise fares while the Bridge users were loudly proclaiming that they were too high.

The Authority met the challenge by carefully husbanding funds and expanding promotional activities. From 1962 on the revenue requirements of the Trust Indenture had not only been met, but surpluses obtained. Redemption of bonds began in 1965. They were purchased in the open market at substantial savings. Thus, bond retirement is seven or eight years ahead of schedule. It is expected the $40,000,000 in bonds outstanding will be redeemed by 1986.

The Executive Organization Act of 1965 placing the State's many agencies under 20 departments transferred the Authority to the State Highway Department, now the Department of Transportation. The arrangement has worked out satisfactorily. When all the bonds are redeemed and the Authority is dissolved, the Mackinac Bridge will come under the full control of the Department.

The Governor and the Legislature in 1968 joined hands in appropriating $3.5 million annually out of the Motor Vehicle Highway Fund to the Authority in return for which fares would be reduced by 60 per cent, or from $3.75 to $1.50 for passenger cars and proportionate decreases for other classes of vehicles. The act became effective January 1, 1969. Traffic increased 22 per cent during the first year of this new fare schedule. Subsequently, the average annual traffic increase has been about four per cent, the same as it was before the rate reduction.

It is a tribute to the designer and builders of the Bridge that it has required no major repairs since completion. Nevertheless, foot by foot annual inspection by the Consulting Engineers results in a long list of maintenance upkeep operations. The Authority's budget in 1959 was a little over a half million dollars. In 1979 it is a little over $2 million. The operation, maintenance and repair of the Bridge are the responsibilities of the Mackinac Bridge Authority.

Acknowledgements

YOUR PICTURE SELECTOR and caption writer simply could not have completed the job without help—lots of it. First, he thanks most humbly the photographers without whom there would have been no pictures. The humility arises from the fact that not only are they highly qualified picture takers, but they are equally courageous. Whenever bridge workers are shown hanging by their figurative fingertips from some lofty perch, the photographer was similarly perched, or more precariously, because he generally needed one hand to hold his camera and the other to operate it. So, hats off to Herm Ellis, boss photographer for the Mackinac Bridge Authority, Harold Bell, worthy assistant, and Mickey Duggan, erstwhile worthy assistant.

Only TV quiz contestants can remember thousands of facts and figures and win thousands of dollars by reciting them. I never won a dime for the ability to remember and recite and, consequently, relied upon a couple of men who helped make the facts and figures, had them on hand, or at least knew where to find them quickly. Thus, we are deeply indebted to John Kinney, Resident Engineer for Dr. David B. Steinman, the designer of the Mackinac Bridge, and his office engineer, Russell H. Garrard, and C. E. Haltenhoff, former Project Engineer for Merritt-Chapman and Scott and now General Manager of the Mackinac Bridge.

I must thank Governor G. Mennen Williams, Prentiss M. Brown, Frank Woodford and Dr. Steinman for their contributions to the book. Without Prentiss Brown's persistence, statesmanship and financial wizardry and without Dr. Steinman's skill and artistry as a bridge builder, we would have no Mighty Mac. Both sacrificed their health and wealth to the success of the project and risked their reputations on its outcome. Dr. Steinman has a poet's feeling about it. Prentiss Brown—well, he's the father.

My thanks go, also, to William Bostick, the designer of the book, for the skill and labor of getting the pictures into proper shape for the printer. And we are indebted to the United States Lake Survey, Corps of Engineers for permission to use their map for the end papers.

Of course, in back of every executive who gets anything down on paper that comes out legible is a secretary, and our office is no exception. In fact, Mrs. Margaret Halava has been correcting our errors, proofreading our letters and generally keeping things running smoothly for us for more than four years. She has risen to the very heights of her usefulness in assisting us with this manuscript. Miss Patricia Peach, who works for Herm Ellis, rode herd on both Ellis and Bell in getting them to produce the prints required for this volume.

I am indebted, finally—really primarily—to the Mackinac Bridge Authority, particularly to Prentiss M. Brown, Chairman of the Authority, for the opportunity as Executive Secretary of the Authority to do this volume.

Mighty Mac, our title, is the term which many of the builders used and which has become common in the Straits area and the Authority has adopted it. It seems appropriate.

LAWRENCE A. RUBIN

Contents

The ruins of Old Fort Mackinac as seen in 1820 by Captain Seth Eastman, U. S. Army. This fort stood on the south side of the Straits near what is now the Mackinaw City end of the Mackinac Bridge. It was the scene of the massacre of most of the British garrison by the Indians in 1763. In 1781 the English established a new fort on Mackinac Island and the old fort deteriorated.

From an engraving in *Information Respecting the History, Condition and Prospects of the Indian Tribes of The United States* by Henry R. Schoolcraft (1854). Courtesy the Burton Collection, Detroit Public Library.

Introduction

By FRANK B. WOODFORD

EVERY STORY with a happy ending begins with
once-upon-a-time. The story of the Mackinac Bridge, linking Michigan's two peninsulas, lifting its tracery of steel and stone above the turbulent waters which join
Lakes Huron and Michigan, is the culmination of a dream with a most happy
ending. Our story's once-upon-a-time goes back into the mists of antiquity, back
through more eons than the mind of the ordinary mortal can comprehend. Shall we
say two billion years?

One of Michigan's most distinguished historians has said that the physical
outlines of the state have been that long in the making. For two million millenniums,
give or take a few months, nature has been at work, creating, destroying, re-making,
shaping, erasing and re-shaping until, satisfied with her work, she has produced
the masterpiece which is Michigan.[1]

Nature is a fickle old wench, full of mischief. She endowed our land with
resources rich beyond count. In the hills she hid iron and copper and coal. She
covered the earth with a garment of trees—forests of majestic pine and noble hard-
wood. She gave rivers and lakes in which the waters run sweet and in which fish
abound. Then, by capricious gesture, she split the face of the land asunder and made,
in reality, two lands, separate from each other. At Michilimackinac, she created a
barrier between these two lands which for time immeasurable prevented easy
access from one to the other.

Let the geologists tell us how that happened.

About a million years ago when Michigan already was old, nature began to
toy with the climate. She created climactic cycles in steady succession, in which
tropical heat was followed by Arctic cold. Each cycle endured for hundreds or
thousands of centuries.

Great masses of ice, glaciers, formed in the polar regions, moved south slowly,
inexorably, and covered the face of the earth. Then these glaciers receded, only to
move down again—and again to draw back. The last of them is supposed to have
retreated about twelve thousand years ago. In its lumbering, heavy-footed progress,
it ground mountains into stubby hills. It strewed boulders across the plains. It gouged
out river courses and dug deep basins into which its melting waters poured, creating
the Great Lakes.

In its passing, the glacier crushed the land link which once joined the Upper and Lower Peninsulas, carving out a trough in the bed rock. Into this trough rushed the waters, and it became the Straits of Mackinac. And so, instead of a bridge of land joining Upper and Lower Michigan, there was left a barrier.

That frivolous and inconsiderate prank of nature was not corrected, despite all man's need and ingenuity, until an epic effort of the mid-twentieth century drew a span across the water and made the two lands one.

The bridge became man's great triumph over nature.

The barrier—the deep, blue-green waters of the Straits—plagued mankind long before the first white man gazed with awe upon its majestic beauty. It is known that the copper of the north was greatly prized by the aborigines, the ancestors or forerunners of the Indians of history. From earliest times there was traffic in copper. The people of the north dug and refined it, and maybe shaped it, and traded it to others. Artifacts of aboriginal copper have been found in many distant places in what is now the continental United States.[2]

To transport that mineral could have been no easy task. The trails through the forest ended at the water's edge—at the great barrier—on either side of the Straits. Venturesome souls must have found it necessary to launch their dugouts or canoes or rafts and hazard the winds and currents in order to carry their treasure from one peninsula to the other. The alternative was wide overland detours through what is Wisconsin and Canada. Yet, somehow, the dangers were surmounted by courage and necessity, and a primitive commercial traffic was carried on.

Because the barrier existed and interposed its difficulties, we can speculate that the value of the copper was increased many-fold. What otherwise would have been a common commodity, easily available to all, became a rare and highly prized, perhaps a sacred, substance. This was partially due to the four-mile stretch of water between the Upper and Lower Peninsulas.

In time—and now we enter the era of measurable history—the Indians took the place of the prehistoric people. Algonquins and Iroquois knew the region well, and eventually it became a battleground over which those two groups fought. The Algonquins were dominant in the north Michigan area. The Chippewas or Ojibways, for the most part, lived north of the Straits; the Ottawas south of it. It was from these people that the entire region, the waters of the Straits, their islands and the land bordering both sides of the barrier, got its name. They called it Michilimackinac, which, in the Algonquin tongue, means great road of departure or, to put it in our own vernacular, the jumping-off place.[3]

For centuries that was what it was to the Indian and the white man who followed him — the jumping-off place. It was the end of the trail, the place where the traveler, coming up against nature's obstacle, surmounted it with difficulty or, as was more than likely, re-oriented himself and changed direction.

A few years ago, a noted archaeologist, studying the ancient Indian trails of Michigan, found their traces still discernible in the beaten earth of forest and field.

He marked those trails on the map of the modern state. His red lines indicated the highways of the red man and showed the principal north-south trails in both peninsulas. In each instance, those red lines end at the Straits of Mackinac, at the water's edge. There are no lines across the water, no indication of island hopping, although, of course, there may have been some. The trails met the shore and went no further. Their continuity was interrupted by the great barrier.[4]

It must be remembered that many of our modern principal highways follow generally the same paths which the Indians marked out. After white civilization came to Michigan, the gravel, tarvia and concrete trails of the white man terminated at the beaches as did those of the Indian.

In their migrations and tribal wars, the Indians moved, because of the Straits, primarily in an east-west rather than a north-south direction. The Iroquois, raiding west out of the St. Lawrence Valley and the lower Great Lakes region, drove the Chippewas from their homes. They fled westward toward what is now Wisconsin. The land trails across the Upper Peninsula and the water routes through the Straits and along the south shore of Lake Superior were their communication lines. These had never been established and could not be maintained across the Straits, and so the fleeing tribes found no refuge in the Lower Peninsula. It was the barrier which stopped them and which decreed that for more than three hundred years for which history keeps account, traffic at Michilimackinac, both Indian and white, was primarily on east-west lines.

When the white men came to Michilimackinac, the pattern of movement followed that set by the Indians. It was a long time before there was penetration of the Lower Peninsula or north-south movement across the Straits.

The period of white or European occupation can be roughly divided into two eras—exploration and exploitation. The line which set the two apart is not always clearly discernible.

As far as recorded history is concerned, the first white man to see Michili-mackinac was the Frenchman, Jean Nicolet. Samuel de Champlain, the governor of New France in the early years of the seventeenth century, did not know the extent of the domain over which he ruled. To find out, he sent forth exploring parties. One of them was led by Nicolet, whose mission was to determine if the upper lakes were, in reality, the roadway to Cathay. It was hoped that the Great Lakes would prove to be the much sought Northwest Passage.[5]

Nicolet set out from Montreal in 1634. He followed the Ottawa River, passing through Lake Nipissing and down the French River into Georgian Bay. For years this was to be the main route of the French and British fur trade. But Nicolet was not seeking furs. He was intent on reaching China. To be prepared for a proper reception by the emperor, he carried an outfit of mandarin robes with him. Nicolet left nothing to chance.

He had been preceded into the Upper Peninsula, probably in 1618, by Etienne Brulé and a young man named Grenoble. This pair found their way to the rapids of the St. Marys River and pushed on to discover Lake Superior. Nicolet chose a different track. He went on through the Straits of Mackinac and became the first known white man to look upon Lake Michigan. He probably followed the south coast of the Upper Peninsula. He arrived, at last, not in China but at Green Bay and was met by half-naked savages. He fired his pistols into the air for their edification and made a good impression. Whether he threw away his mandarin costume, or whether, possessing a sense of humor, he draped it over an Indian, is something the history books fail to mention. The point is that Nicolet passed through the Straits, and there is no hint that he crossed it even to examine the Lower Peninsula.

Only the Upper Peninsula was known for thirty-five years after Nicolet's passage. Not until 1669 was the Lower Peninsula visited by Europeans. The Straits, in other words, was a passage, not a crossing place.

Having discovered the Michigan north country, the French devoted themselves to its exploitation. They found it yielded two crops, both much sought after. These were furs and human souls. While the traders and *coureurs de bois* ranged the forests to get the beaver pelts, the Black Robes, the fathers of the Jesuit and Recollet orders, followed in their footsteps or struck out on their own, seeking converts among the savages. One of these was Father Jacques Marquette. In 1671 he established a mission which he called St. Ignace. By that act, civilization came to the Straits.

In the years immediately following the founding of St. Ignace, there was considerable traffic through the Straits. Marquette, himself, accompanied by a young woodsman named Joliet, went through on a journey which led to the discovery of the upper Mississippi. Marquette died before his return trip was completed, but his remains eventually found eternal repose under the altar of his church at St. Ignace.

La Salle passed through in 1679 on a venture marked by ill-fortune, for the ship he sailed, the *Griffin,* which was the first to appear on the Great Lakes, foundered on its return from Green Bay with the loss of all hands and a cargo of fur. There are some who speculate that the bones of the *Griffin* may lie on some reef or bar not far from Mackinac.

As with the earlier expeditions, so was it with the later. The route was from east to west and return, never from north to south, and while small colonies of settlers established themselves at various places in the Upper Peninsula, the lower part of Michigan remained a dark continent. Those from the north who went south at all, did so by water through Lake Huron. There was no passing the barrier.

In 1690, the Sieur de Louvigny built Fort de Buade at St. Ignace on the point of land jutting into the Straits of Mackinac where Marquette had placed his mission some years before. There it stood guarding the passage until the Church prevailed upon the government to limit the fur trade and abandon the army posts. The missionaries complained that the traders with their brandy and their loose habits were corrupting the Indians and making the salvaging of souls an all but impossible task. Obeying the edict, Fort de Buade was evacuated in 1698 by Antoine Cadillac, who, three years later, persuaded the authorities to let him found a new settlement at Detroit. For a long time before and after these events, Michilimackinac, according to a modern historian, "was the most important place in the west."[6]

It was too important to stand abandoned for long. When the British extended their trade and influence into the upper Great Lakes region, the French again built forts to hold them back. Michilimackinac was a key, a strategic place. It controlled the lines of communication between the fur producing areas of the west and the markets in the east. So, in 1715 a new stronghold was built. This time the barrier was breached ever so slightly. Fort Mackinac was placed on the south shore of the Straits where Mackinaw City stands today.

In time, the British prevailed and all of New France passed to British control, Fort Mackinac included. On September 28, 1761, it was occupied by Lieutenant William Leslye and a small detachment of the green-coated Royal American Regiment. During Pontiac's rebellion of 1763, the fort was captured and most of the garrison was massacred by the Chippewas. Fort Mackinac was re-occupied by the British in 1766, but in 1781, prompted by fear of the rebelling American colonists, the commandant, Major Patrick Sinclair, moved his installation to Mackinac Island, which henceforth would be the citadel of the north. It remained in British hands until 1796 when, along with the rest of Michigan, it was handed over to the United States. For the next forty years, Mackinac Island was the center of the fur trade of the northwest and the seat of operations of John Jacob Astor and his American Fur Company. During this period, Michilimackinac attained great economic importance.

All the while, from the beginning of French exploitation almost to the mid-nineteenth century, fur was king and millions of dollars worth of it passed through the Straits. It was a rich, heavy and continuing commerce. The Indians and white trappers ranged far into Canada, Wisconsin and Minnesota. Mackinac Island was the focal point where expeditions were outfitted and whence they set out, and the place to which the heavily laden canoes and batteaux returned. The flow of trade goods and pelts moved from east to west, and west to east. There was trapping, of course, in the Lower Peninsula; a few posts and missions were established, and in the extreme southern part, communities flourished. But not much traffic crossed from Mackinaw Point or from St. Ignace to Mackinac Island. Most of it was through the narrow waters of the Straits. Generally speaking, there were no white men in the Lower Peninsula north of Saginaw Bay, and, except for a few isolated settlements along the shores of the lakes, the Upper Peninsula was virtually unknown.

Not until after the War of 1812 was there a serious, organized attempt to penetrate the curtain of darkness which enshrouded the Michigan north country. In 1805 Michigan became a territory, but only the eastern half of the Upper Peninsula was included in it. The rest was attached to Indian Territory until 1818, when Michigan Territory was greatly enlarged to take in all of the present state as well as Wisconsin and part of Minnesota.

About that time the first movement of American settlers into Michigan began. Settlement was slow at first because no one had a very clear idea of what lay within her boundaries. Moreover, government surveyors had given Michigan a bad name. They characterized it as a land of sand and swamp, unfit for cultivation. Lewis Cass, governor of the Territory, tried to convince the federal government that those reports were calumnies. What was needed, he insisted, was scientific investigation to determine the truth and place before the world the true picture of Michigan so settlers would not by-pass the Territory in favor of Indiana, Illinois and points beyond.[7]

An expedition was authorized by the War Department in 1820 with Cass in command of a group of competent scientists, engineers and observers. The group traveled from Detroit in a flotilla of canoes, and the first major stop was at Mackinac Island.

The subsequent reports of the various members confirmed what thousands of visitors to Michilimackinac have discovered since. They were left breathless by the grandeur and beauty of the Straits area. One hard-bitten officer of the military escort grew lyrical in his description of the vistas which unfolded before him as "sublime views of a most illimitable and magnificent water prospect."

The most articulate member of the expedition was Henry Rowe Schoolcraft, whose name was to be permanently linked with the Upper Peninsula. No detail of flora, fauna, rock or mineral escaped his inquiring eye. He described Mackinac Island—and his words applied equally to the whole Straits region—as a place "celebrated for the salubrity of its atmosphere." Generations of hay fever sufferers have confirmed that statement. Completely charmed by what he found, he wrote in glowing words the first literature to awaken the interest and enthusiasm of the tourist. He found the island to have a settlement of one hundred and fifty houses, some of which were handsomely painted. In his journal, he dwelt on the history of the Straits country; he noted the abundance of white fish, which, before long, would provide a thriving fisheries industry in the upper lakes and which, along with other game and fish, would make northern Michigan a sportsman's paradise.[8]

Schoolcraft studied the geology of the area, and his findings accurately coincided with those of the bridge planners one hundred and thirty years later. He noted particularly the cavities in the rock which were to be heatedly and erroneously offered as a reason why the weight of the bridge could not be supported.

Schoolcraft also paid attention to the commerce of the Straits, describing the mats, baskets, moccasins, shot pouches and maple sugar of Indian manufacture, and he itemized the volume of furs which annually passed through Mackinac. He referred to the historic importance of the passage through the Straits of the first steamship, the *Walk-in-the-Water*, in 1819. With some of his companions, he visited the tip of the Lower Peninsula and remarked upon the difficulty, due to winds and current, of crossing the four-mile-wide barrier.

Cass's party left Mackinac, passed through the St. Mary's River, and proceeded north and northwest. They found evidences of copper in the Upper Peninsula near Ontonagon, and in succeeding expeditions, Douglass Houghton, in 1841, confirmed the richness and extent of those deposits. In 1844, William A. Burt discovered a treasure trove of iron ore. These discoveries, and the wide public interest they aroused, led to the development of the mining industry. That, in turn, prompted the building of the Sault Ste. Marie Canal and locks in 1855. About the time of the Civil War, lumbering became a major enterprise in both the Upper and Lower Peninsulas, and the exploitation of these resources led to the settlement of the land on both sides of the Straits.

There was one portentous event between the years when American sovereignty was established over Michigan and those in which the mineral and lumber wealth began to flow from the Upper Peninsula mines and forests. That was the political welding of the two peninsulas into a state.

In the 1830's, Lower Michigan had grown sufficiently in population and advanced far enough economically to qualify the Territory for statehood. But there was a stumbling block which prevented a smooth political transition. That was a dispute with Ohio over possession of a seven-mile wide strip—the so-called Toledo strip. Ohio claimed it. So did Michigan, which had possession of it. Until this boundary question could be settled amicably, Michigan could not be admitted to the Union.

War threatened between Michigan and Ohio; both sides went so far as to mobilize their militias. Then cooler heads intervened and the federal government proposed a compromise. The disputed Toledo strip would be given to Ohio, and in return Michigan would get all of the peninsula lying north of the Straits.

Most people in southern Michigan were inclined at first to reject this offer. Compared to what they would have to give up to Ohio, they could see no value received. One Michigan official sardonically remarked that Congress intended "to give us a strip of country along the south shore of Lake Superior, where we can raise our own Indians in all time to come and supply ourselves now and then with a little bear meat for delicacy."[9]

This about summed up the general Michigan attitude. It stemmed from the isolation of the Upper Peninsula, from its inaccessibility resulting, in part, from the barrier at the Straits. But such men as Cass and Schoolcraft, possessing first-hand knowledge of the potential worth of the north country, vigorously advocated the substitution of the Upper Peninsula for the Toledo strip, and it was predicted in what proved to be a masterpiece of underappraisal that in time the Upper Peninsula would have a value of forty million dollars.

In the end, the advocates of compromise prevailed, and on January 26, 1837, President Andrew Jackson signed the bill which made Michigan a state—a state composed of two peninsulas. For a long time thereafter, many citizens of the southern part of the state believed Michigan had come out on the short end of the deal. For a long time the geographical separation caused by the Straits resulted in a feeling of economic and political separateness instead of the natural existence of a single political community. So strong was the feeling that there was sentiment, still prevailing only a few years ago, that the Upper Peninsula would fare better if it was attached to Wisconsin or, preferably, allowed to become a separate state.

The lack of easy access between the two parts of Michigan was responsible for the persistence of this feeling. During periods of the year when navigation was closed, it was virtually impossible to travel from one peninsula to the other. Members of the state legislature representing northern constituencies found it necessary to go to Lansing by way of Wisconsin and Illinois, and sometimes to spend the entire winter away from their homes. When the railroads came, it often was easier to go to Chicago than Detroit from many Upper Peninsula points. As a result, much of the northern trade and commerce was oriented to Minneapolis, Milwaukee and Chicago, instead of to Lower Peninsula cities. Lower Peninsula people regarded it as the crowning treason when many of their northern cousins gave their loyalty to the Chicago Cubs rather than the Detroit Tigers.[10]

The barrier exacted a heavy toll on business and cordial human relations to the disadvantage of the whole state. The effect on the economic development of the Upper Peninsula was particularly harmful.

Of course there was across-the-Straits traffic of sorts during all of those early years. The occasional Indian in his canoe undoubtedly found reason to make the trip. White men, too, sailed or paddled across from one peninsula to the other; it is known that the garrisons on Mackinac Island took barges to the mainland for loads of firewood. When winter locked the Straits, there was passage back and forth over the ice, on foot, on horseback, by sleigh and even by dog team. But that was hazardous business. Those who read the Sunday supplements not so many years ago can recall the perennial story of the heroic mail carrier who braved blizzard and icy blast and of the other adventurous travelers who sometimes did not make it over the ice.[11]

None of these early means of transportation could accommodate any but light, infrequent traffic. For the general public there was but little linkage during the summer months and practically none at all in winter.

The first regular ferry service was established in 1881, but it was limited and irregular and operative only in favorable weather. In the 1880's, the railroads reached the Straits and, like the highways and Indian trails, stopped abruptly at the water's edge. The necessity for continuation of rail service led to the first real linkage when a car ferry was built and put into operation from Mackinaw City to St. Ignace. While this and vessels of similar type handled the railroad business and served the public in a limited way, there was felt a need for something better.

In 1883, the Brooklyn Bridge was completed. The imagination of the American people was stimulated and new hopes were awakened in the minds of northern Michiganders. William Saulson, a merchant with stores at St. Ignace and Seney, in 1884 published newspaper advertisements carrying pictures of the Brooklyn Bridge and a caption which read: "A Glimpse of the Future—Proposed Bridge Across the Straits of Mackinac." The same year, the *Grand Traverse Herald* of Traverse City stated that year-round ferry service was a failure, and the editor declared that a bridge or tunnel was necessary if a great east-west route ever was to be established through Michigan. The *Grand Traverse Herald* article was re-printed in other Michigan newspapers and evoked considerable discussion.[12]

Meanwhile, the Straits area was becoming increasingly important as a tourist and vacation region. Each summer, the railroads and their affiliated steamship lines brought throngs of visitors to Mackinac Island. In 1887, a group of men with railroad interests, headed by Commodore Cornelius Vanderbilt, opened the famous Grand Hotel on the island. Vanderbilt remarked at that time, "We now have the largest well-equipped hotel of its kind in the world for a short season business. Now what we need is a bridge across the Straits."[13]

Not much more was to be said about such a project for nearly fifty years. It took the automobile to revive interest and bring reality out of the dream.

By the time the United States had become involved in World War I, the automobile was well on its way to becoming the nation's foremost means of transportation, and touring in the family car was getting to be the number one outdoor sport of thousands. To stimulate interest in touring and promote the cause of good roads, local automobile clubs in association with the East Michigan Pike Association sponsored annual motor cavalcades from Detroit to Sault Ste. Marie. The first of them was run in 1916. They served a most useful purpose in stimulating a demand for better highways, and by 1920 Michigan was embarked on a huge road building program. Like the Indian trails, the mud roads and the rail lines before them, the improved trunk line highways also came to dead ends at the Straits of Mackinac.[14]

Responding to the clamor of motorists for across-the-Straits transportation, the railroad ferries provided limited facilities for carrying automobiles back and forth. The service may have been good enough for those early days, but it was irregular and the cost was burdensome. U. S. Senator Prentiss M. Brown, of St. Ignace, the father of the Mackinac Bridge, recalled that in those days it cost him sixteen dollars to ship his Model T Ford across the Straits on a car ferry. In winter, it was a particularly difficult trip, and while later ferries were designed as ice breakers, they sometimes became locked in the ice for considerable periods of time.

Before long the volume of automobile traffic had grown substantially, and a public demand for better facilities was heard. Horatio "Good Roads" Earle, Michigan's first highway commissioner, talked about a "floating" tunnel across the Straits in 1920. The following year, a New York engineer, Charles Evan Fowler, proposed a bridge which would leap-frog the islands in the Straits.[15] In 1923, the state instituted its own ferry service, primarily for automobiles. Ultimately, the state, through its highway department, was to purchase or build a total of eight fine vessels to handle the traffic. Unfortunately, while these ferries were able to operate on fairly regular year-round schedules, there were seasons in which they fell far short of meeting the heavy demands placed on them. During hunting season, it was not unusual for automobiles to be lined up for five or more miles, waiting passage. A wait of several hours was common.

Naturally, all this led to a revival of the demand for a bridge. As early as 1933, engineers were seriously studying the possibility. The next year a bridge authority was created.[16] It really came into being in order to sponsor a depression era bid for federal financing. A total cost of $35,000,000 was the estimate, and a government loan of 70 per cent of that sum was applied for. The request was rejected, and it became a matter of general agreement that if a bridge was to be built, it would have to be the state that would build it.[17]

As matter of fact, tangible steps were taken by the predecessor to the present Mackinac Bridge Authority. Under the chairmanship of Peter Stackpole, of Detroit, a railroad executive, and his successor, former State Highway Commissioner G. Donald Kennedy, actual work was started. The engineering firm of Modjeski and Masters was retained, and preliminary plans were drawn. A new location was selected in preference to those previously considered for the bridge, and, perhaps most important of all, a causeway was constructed in 1941 on the north side of the Straits at a cost of about $700,000. The new location was adopted by the present Authority, and the causeway was used. This preliminary work was said to have resulted in savings of more than $3,000,000 when the present bridge was built. It also has been stated that plans were sufficiently advanced in 1941 so that had World War II not interfered, construction of a bridge undoubtedly would have been undertaken at that time.

For a considerable period preceding World War II, and for a time following, there was heated debate on the merits of the project, from the standpoint of construction practicability and necessity, with discussion of financing. Public and official opinion was sharply divided on the subject of a bridge, and it became necessary to do a good deal of educating. Competent engineers said a bridge could not be built. The honeycombed rock which Schoolcraft observed would not support such a structure. Neither, it was declared, could a span of the height and length required withstand the winter gales, the pressure of the currents and the force of the ice. Last of all, it was claimed, traffic would never be sufficient to pay the capital costs and the financing charges which a bond issue would entail, and eventually it would become a direct burden upon the taxpayers.

Replies to each of these objections were made by equally competent authorities. One of the latter was the late James H. Cissel, Professor of Engineering at the University of Michigan. In a notable address on the subject of the proposed bridge in 1937, he endeavored to establish the feasibility of the project from every consideration. He concluded with these words:

"Recent advances in engineering and construction now make it possible for Michigan to overcome the physical defect of her divided peninsulas and by this proposed link be welded into a single individual state."[18]

Successful completion of the San Francisco Golden Gate Bridge, a span comparable to that visualized for the Straits of Mackinac and involving many identical engineering and financing problems, helped to weaken the arguments of the opposition.

One of the most eloquent and respected voices raised in behalf of the Straits Bridge was that of Chase S. Osborn, Nestor of the North and one-time governor of Michigan. Attainment of a venerable age did not dim Osborn's spirit of pioneering adventure.

At one period, he wrote in 1935, he had opposed the bridge, but his ideas had changed largely due to the influence of Dean Mortimer E. Cooley, of the University of Michigan College of Engineering. There was no basis, he declared, for the argument that financial difficulties were too great an obstacle.

"Suppose," he said, "on a trunk line—and the Straits road is that—there was a mud hole or chasm or abyss or sink-hole eight miles wide that every car had to be pulled over or across or through. Something would be done about that at once . . . ."

And then he went on to blast the proposition that the Upper Peninsula and the northern part of the Lower Peninsula had reached their limits of economic development.

"Michigan is unifying itself," he wrote, "and a magnificent new route through Michigan to Lake Superior and the northwest United States is developing, via the Straits of Mackinac. It cannot continue to grow as it ought with clumsy and inadequate ferries for ANY portion of the year."[19]

The effect of the bridge upon potential tourist business—Michigan's second most important industry—was not overlooked by the proponents of the project, and this undoubtedly was as strong a consideration in influencing public opinion as any arguments advanced.

Gradually, the chorus swelled with such powerful, influential voices as those of Senator Brown; George Osborn, publisher of the *Sault Ste. Marie Evening News* and son of Chase S. Osborn; former Highway Commissioner Murray D. Van Wagoner; and W. Stewart Woodfill, Manager of the Grand Hotel. And there were many more. The opposition (a better word would be misunderstanding) was still vocal, but little by little it was becoming muted.

Much of the credit for convincing the legislature to take favorable action belongs to Woodfill. At the end of the 1949 season he closed his Mackinac Island hotel and went to Lansing. There he spent the entire winter session of 1950, bringing all of his enthusiasm and powers of argument to bear on hesitant and uninformed legislators. His own account of that winter is a real saga of dedicated persuasion. Finally, he broke down the last resistance, and in 1950 the legislature created a new Mackinac Bridge Authority and appropriated token funds to pay for preliminary studies and surveys.[20] On June 6, the members of the Authority, appointed by Governor G. Mennen Williams, took office. They were Fred W. Zeder (succeeded after his death by Mead L. Bricker), Charles T. Fisher, Jr. (succeeded after his death by his widow), George Osborn, Murray D. Van Wagoner, W. J. Cochran, Highway Commissioner Charles M. Ziegler, and Senator Brown. Lawrence A. Rubin was named executive secretary.[21]

Now, with the Authority established, and after exploratory reports and surveys had been completed, there was one last step before starting actual bridge construction. That was to arrange the financing. Senator Brown and his colleagues labored diligently at that assignment during 1952 and 1953, and on December 17 of 1953, bids were accepted from a group of underwriters for the sale of $99,800,000 worth of Mackinac Bridge Authority bonds. Two months later, the bonds were delivered, the cash was received, the first construction contracts were awarded, and work was started under the general supervision of David B. Steinman, a noted bridge engineer with the soul of a poet.

The rest is a familiar story. The people of Michigan watched, as did the world, the remarkably rapid progress as the piers were sunk, the towers were raised, and the superstructure was flung across the waters which once had known only the frail birch bark canoes of the Chippewas and Ottawas.

It is no anti-climax to state that the bridge was finished in 1957 and was open for use on November 1 and was dedicated with appropriate ceremony on June 28, 1958.

The dream at last was a reality of steel and stone.

Michigan's two great peninsulas were linked and the state was united.

The barrier was forever broken!

REFERENCES

1. F. Clever Bald, *Michigan in Four Centuries* (New York, 1954), p. 3.

2. Luke Scheer, *Michigan and the Old Northwest* (Detroit, 1945), p. 9.

3. *Sault Ste. Marie Evening News,* November 1, 1957, an article by Emerson F. Greeman of the Michigan Historical Commission. There are several versions of the meaning of Michilimackinac. Bald points out that the term *Michilimackinac* was given to the entire Straits area, not just to the island.

4. Walter B. Hinsdale, *Archaeological Atlas of Michigan* (Ann Arbor, 1931).

5. Bald, *op. cit.,* pp. 24-68; see also Thomas McIntyre Cooley, *Michigan, a History of Governments* (Boston, 1885), pp. 2-19.

6. Bald, *op. cit.,* p. 45.

7. Frank B. Woodford, *Lewis Cass—The Last Jeffersonian* (New Brunswick, 1950), pp. 126-131.

8. Henry Rowe Schoolcraft, *Narrative Journal of Travels ... 1820,* ed. Mentor L. Williams (East Lansing, 1953), pp. 78-88, 267-8, 368.

9. Lawton T. Hemans, *Life and Times of Stevens Thomson Mason* (Lansing, 1920), pp. 195-200.

10. James H. Cissel, "Bridging the Straits of Mackinac," *Michigan Alumnus Quarterly Review,* XLIII (Spring, 1937), 530-537.

11. Prentiss M. Brown, "The Mackinac Straits Bridge," a clipping from an unidentified publication in the library of the *Detroit Free Press.*

12. *Sault Ste. Marie Evening News,* November 1, 1957, an article by A. F. Mahan, Jr. See also *The Story of the Mackinac Bridge* (anonymous) (Mackinaw City, 1955).

13. *Ibid.*

14. Frank B. Woodford, *We Never Drive Alone* (Lansing, 1958), pp. 45-48.

15. Cissel, *op. cit.*

16. This authority, like a second one, became inoperative and expired under the law.

17. Prentiss M. Brown, *The Mackinac Bridge Story* (Detroit, 1956), pp. 4-6.

18. James H. Cissel, radio address, Station WJR, Detroit, December 17, 1937. The original script is in the possession of the Mackinac Bridge Authority, St. Ignace.

19. Chase S. Osborn in a letter to the *Mining Journal,* Marquette, December 21, 1935. Copy in the possession of the Mackinac Bridge Authority, St. Ignace.

20. *Sault Ste. Marie Evening News,* November 1, 1957. Mr. Woodfill, in an interview, gave a detailed account of his legislative activities.

21. Mr. Ziegler's membership was ex officio. He was automatically succeeded July 1, 1957 by John C. Mackie when the latter became Highway Commissioner.

The Design of The Mackinac Bridge

By D. B. STEINMAN

THE MACKINAC BRIDGE is the greatest bridge in the world. Its cost is more than that of the George Washington Bridge and the Golden Gate Bridge combined. The record cost of $99,800,000 is a measure of the magnitude and the difficulty of the project. Both artistically and scientifically, it is outstanding. No effort has been spared to make it the finest, safest, and most beautiful bridge the world has ever seen.

The bridge is five miles long. In the middle of the bridge, in the deepest water, spanning a submerged glacial gorge, we have a suspension bridge bigger than the George Washington Bridge. With a length of 8614 feet from anchorage to anchorage, Mackinac is the longest suspension bridge in the world.

One day during the building of the bridge, Grover Denny, the genial and resourceful construction superintendent on the foundation work, said to me, "Doctor, I believe that you have made an important mistake in a decimal point." Startled by this remark, I asked: "What do you mean, Grover?" To which he replied: "Doctor, you have been telling people that this bridge is good for a century. But I want to go on record as saying that this bridge will be standing a thousand years from now!"

Before construction, people said that the rock underlying the Straits, because of the unusual geological formation, could not possibly support the weight of a bridge. To resolve any doubts, outstanding geologists and soil-mechanics authorities were consulted. Exhaustive geological studies, laboratory tests, and "in-place" load tests on the rock under water at the site established, without a doubt, that even the weakest rock under the Straits could safely support more than 60 tons per square foot. This is four or more times greater than the greatest possible load that would be imposed on the rock by the structure, including the combination of dead load, live load, wind load, and ice pressure. The foundations were made large enough and massive enough to keep the maximum possible resultant pressure under 15 tons per square foot on the underlying rock.

The geologists, moreover, pointed out that during the past one million years the rock under Mackinac Straits has been preloaded and pretested by the weight of a solid glacier, from one to five miles high, representing a load ten to fifty times greater than the maximum pressure under the bridge foundations.

The foundations of the Mackinac Bridge contain more than a million tons of concrete and steel to form the massive piers and anchorages. Three-quarters of this mass is under water, to provide enduring stability. These foundations will be more enduring than the pyramids.

Because the public had been alarmed by unscientific claims that no structure built by man could withstand the ice pressure at the Straits, I added a further generous margin of safety. According to the most recent engineering literature on the subject, the maximum ice pressure ever obtained in the field is 21,000 pounds per lineal foot of pier width. In the laboratory, under specially controlled conditions for maximum pressure, the greatest ice pressure producible is slightly greater, 23,000 pounds per lineal foot. I multiplied this higher figure by five and designed the piers to be ultra-safe for a hypothetical, impossible ice pressure of 115,000 pounds per lineal foot.

With the maximum possible ice pressure multiplied by five, and the safe foundation pressure divided by four as a basis for design, the combined factor of safety is 20 for the ultra-safe design of the piers against any possible ice pressure. For still further safety against any possibility of ice damage, the concrete of the piers and anchorages is protected by inter-locking steel sheet piling, steel caissons, and wrought iron armor plate.

The maximum pull of the two cables on the anchorage is 30,000 tons. Each anchorage has a mass of 180,000 tons of concrete and steel, providing a generous safety factor of six against the maximum cable pull.

The public had been irresponsibly told that no structure could resist the force of storms at the Straits. So I made the design ultra-safe against wind pressure, too. The greatest wind velocity ever recorded in the vicinity was 78 miles per hour; this represents a wind of 20 pounds per square foot. I multiplied this force by two and a half, and I designed the bridge to be ultra-safe against a hypothetical, unprecedented wind pressure of 50 pounds per square foot.

The crowning feature in the design of the Mackinac Bridge for ultra-safety is its unprecedented achievement of perfect aerodynamic stability. This means absolute security against the initiation or amplification of oscillations which are due to wind, the kind of oscillations ("galloping" and "twisting") that destroyed the Tacoma Narrows Bridge in 1940.

The main span at Mackinac is a suspension bridge, which is inherently the safest possible type of bridge. The stiffening trusses are 38 feet deep, or one-hundredth of the span length. This is 68 percent greater than the corresponding ratio of the Golden Gate Bridge. But even without this generously high depth-ratio, the Mackinac suspension span would have more than ample aerodynamic stability. In fact, by utilizing all of the new knowledge of suspension bridge aerodynamics, particularly my own mathematical and scientific discoveries and inventions, I have made the Mackinac Bridge the most stable suspension bridge, aerodynamically, that has ever been designed.

This result has been achieved not by spending millions of dollars to build up the structure in weight and stiffness to resist the effects, but by designing the cross-section of the span to eliminate the cause of aerodynamic instability. The vertical and torsional aerodynamic forces which tend to produce oscillations have simply been eliminated.

The basic feature of this high degree of aerodynamic stability is the provision of wide open spaces between the stiffening trusses and the outer edges of the roadway. The trusses are spaced 68 feet apart, but the roadway is only 48 feet wide. This leaves open spaces ten feet wide on each side, for the full length of the three suspension spans.

When you drive over the three suspension spans in the central portion of the bridge, you can see the stiffening trusses beyond the edges of the roadway, with the open spaces between them. These open areas constitute the scientific design. They eliminate the closed corners in which pressure concentrations are producible by wind, and they also eliminate the solid areas on which such pressure differences would otherwise act to produce oscillations of the span.

By this feature alone (the open spaces between roadway and stiffening trusses), the critical wind velocity (the wind velocity at which potential oscillations can start) was increased from 40 miles per hour to 632 miles per hour!

But I was not satisfied with raising the critical velocity to this fabulous figure. For still further perfection of the aerodynamic stability, I provided the equivalent of a wide opening in the middle of the roadway on the suspension spans. The two outer lanes, each twelve feet wide, are made solid, and the two inner lanes and the central mall, 24 feet wide together, are made of open-grid construction of the safest, most improved type. By this additional feature of aerodynamic design—adding the central opening to the lateral openings previously described, I achieved a further increase in aerodynamic stability and raised the critical wind velocity from 632 miles per hour to a critical wind velocity of infinity!

These results and conclusions have been independently confirmed by laboratory and wind-tunnel tests on small-scale and large-scale section-models of the bridge. The wind-tunnel tests show conclusively that the Mackinac Bridge has complete and absolute aerodynamic stability against *all* types of oscillations (vertical, torsional, and coupled) at *all* wind velocities and *all* angles of attack of the wind.

The Mackinac Bridge represents the triumph of the new science of suspension bridge aerodynamics. It represents the achievement of a new goal of perfect aerodynamic stability, never before attained or approximated in any prior suspension bridge design. In other large modern suspension bridges, the critical wind velocities range from 30 to 76 miles per hour. In the Mackinac Bridge, the critical wind velocity is infinity.

Michigan's bridge is not only a scientific and economic triumph. It is also an artistic achievement.

Devoted thought and study were applied to the development of forms, lines, and proportions to produce a structure of outstanding beauty. A suspension bridge is a naturally artistic composition, with the graceful cable curves and the symmetry of the three spans, punctuated by the dominant soaring towers and framed between the two massive, powerful anchorages. There is symmetry about each tower, and over-all symmetry of the three-span ensemble. The lofty towers are architectural compositions of vertical and horizontal lines; and each horizontal member is arched for inflection and pierced for artistic interest and distinction. The design is functional and efficient, actually saving in cost while producing an effect of notable beauty. The suspension span framed by the two lofty towers is "a harp outstretched against the sky," "a net outspread to hold the stars." The bridge as a whole is a "symphony in steel and stone," "a poem stretched across the Straits."

For the painting of the bridge, I chose a two-color combination—foliage green for the span and cables and ivory for the towers—to express the difference of function. (I may be joshed about the "ivory towers.")

During the cable stringing, lights were strung along the catwalks for the night work. They were like a necklace of pearls and inspired artists and poets. My suggestion that such illumination of the cables, necessary for construction, be made a permanent installation was enthusiastically adopted. With lights strung along the cables and with flood-lighting at the towers, the beauty of the lines and forms as seen by day will be continued in poetry and magic at night.

To a dedicated bridge designer, a bridge is geometry transmuted into poetry and music.

The Mackinac Bridge is my crowning achievement—the consummation of a lifetime dedicated to my chosen profession of bridge engineering. As far back as 1893, when I was a newsboy selling papers near the Brooklyn Bridge, I told the other newsboys that someday I was going to build bridges like the famous structure that towered majestically above us. They laughed at me. Now I can point to 400 bridges I have built around the world, and to my masterwork—the Mackinac Bridge—the greatest of all. The realization, one after another, of dreams that seemed hopeless leaves me reverent and humble.

To me, the Mackinac Bridge has been a challenge and an opportunity—a challenge to conquer the impossible, to build a bridge that people said couldn't be built; and an opportunity to apply consecrated, scientific, resourceful design to achieve the finest, safest, and most beautiful long-span bridge that the science and art of bridge engineering can create.

The Mackinac Bridge represents a triumph over staggering obstacles and difficulties—some man-made and others imposed by nature. We had to overcome the difficulties of legislation and financing in the face of ignorance, skepticism, and prejudice. And we had to conquer the problems of the challenging natural conditions —the magnitude and depth of the crossing, the unique geology of the straits, and the alleged hazards of ice and tides and storms. The structure has been made generously safe to defy all these natural forces with an unprecedented high margin of safety.

Aerodynamically, it is the safest bridge in the world; in fact, it is the first long-span bridge ever designed and built to have perfect assured aerodynamic stability for all wind velocities up to infinity. This result has never before been approached or attained.

Finally, my staff and I are proud of our record of building the Mackinac Bridge within our estimate of cost and within our estimate of time. We have generously fulfilled all our promises and commitments. We have kept faith with the Bridge Authority and with the people of Michigan. The Mackinac Bridge may well be called *the bridge that faith has built*.

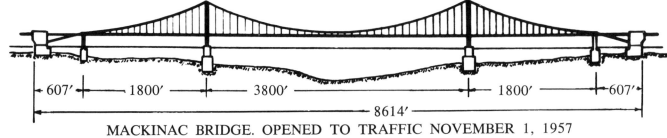

MACKINAC BRIDGE. OPENED TO TRAFFIC NOVEMBER 1, 1957
(Dedication Celebration—June, 1958)

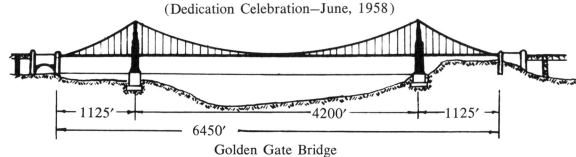

Golden Gate Bridge
(completed 1937)

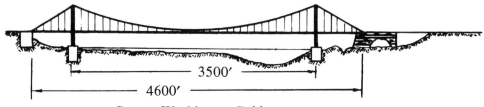

George Washington Bridge
(completed 1931)

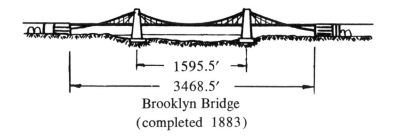

Brooklyn Bridge
(completed 1883)

Comparative Magnitude of World's Great Suspension Bridges

I

Before the Start

1. Since this is the story of the construction of the Mackinac Bridge, it starts with the where-with-all that moved the contractors into action—a certified check for $96,400,033.33. Prentiss M. Brown, Authority Chairman, holds the check, surrounded by Authority members George A. Osborn, Charles T. Fisher, Jr. (deceased), Executive Secretary Lawrence A. Rubin, and members William J. Cochran and Mead L. Bricker. Absent members are Murray D. Van Wagoner and Charles M. Ziegler. Date: February 17, 1954.

2. A close-up of the proceeds of the sale of $99,800,000 worth of bonds. The difference represents the fees paid to the hundreds of firms all over the United States who handled the distribution of the bonds.

A three-picture story dramatizes one of the reasons for a bridge.

3. An aerial shot of the Mackinaw City docks shows some 800 cars waiting for a ferry.

4. A ground view from the road end of the dock. The camper pulling the boat (left foreground) didn't get to St. Ignace until five hours after this picture was taken.

5. Another aerial shot showing ferry crossers lined up on the highway two miles from the dock. Ferry capacity was 460 cars per hour. Bridge capacity is 6,000 cars per hour.

6. Merritt-Chapman and Scott, world's foremost builder of marine foundations, was awarded a $25,735,600 lump sum contract (low bid) for 33 marine foundations. This water-borne forest of derricks represented in 1954 the largest armada of marine equipment ever assembled for a peacetime project.

7. Pinpointing a precise location on water. One of the six triangulation survey towers anchored in the Straits, three on each side of the proposed centerline of the bridge. On land, a surveyor simply drives a stake.

8. For surveying from land, there were seven survey towers at known fixed locations. By sighting along lines from one location to another, and by directing surveyors in boats with walkie-talkies to position themselves where these lines crossed, locations of piers were pin-pointed.

9. A causeway 4200 feet long built in 1941 for a bridge planned then and halted by World War II. Dr. D. B. Steinman, designer of the Mackinac Bridge, used this causeway saving some $3,000,000 in construction costs.

II
Getting a Toe Hold
in the Straits
1954

There are two general methods of marine foundation construction, cofferdam and caisson. Both were used on Mackinac Bridge. Three of the largest piers founded on rock more than 100 feet below the surface were built with caissons. The remaining 30 employed the cofferdam method. Of these, five were cofferdams built on steel piles driven to rock. [Pictures 10 through 26]

MUD

BEDROCK

10. Artist's sketch showing one of the unique features of Mackinac Bridge foundations. All were built from rock up to an elevation of eight feet below the surface excepting the anchorages. From this point up heavily reinforced slender pedestals were erected to support the steel superstructure. This was done to counteract ice, which often moves back and forth through the Straits with the wind. Since the ice is rarely up to three feet thick, it can move around the relatively narrow shafts, while the pressure thereon is taken up by the huge heavy foundations.

11. The two huge foundations in which the cables are anchored measure 115 x 135 feet, about one-third the size of a football field. They are toed into rock more than 90 feet below the surface. The first step in building them required the land construction of three similar steel 106-ton frameworks, 135 x 35 x 75 feet high.

12. The first of the three steel frames was loaded aboard a barge and towed to a position pre-selected by the surveyors.

13. When in position the steel frame was picked off the barge by three marine cranes, sometimes called "whirlies" because they turn a full circle, and gently lowered to the overburden. This is the sand, mud, clay, broken stone and miscellaneous soft material lying under water and above bedrock.

14. With surveyors radioing directions (note tower in background and white sighting rods welded to four sides of steel frame) the crane operators maneuver the frame into precise position within a tenth of an inch.

15. Once in position, the frame is locked to rock by driving huge spud piles (117 pounds to the lineal foot) through slots in the frame, through the overburden, into the bedrock until the pile could no longer be driven. (Every blow was counted and the depth of penetration it caused was precisely measured.)

16. When the first of the three steel frames was tied to rock with spud piles (see five of eight above), the second or middle section of the frame was attached both above the surface and below as a diver descended to perform underwater operation. The procedure was repeated, with the third steel frame completing the entire framework for the foundation.

17. Clam-shell buckets, operated from three busy "whirlies," dig overburden out of cofferdam, which was not dewatered. A jet (center of picture) removes fine materials in the manner of a vacuum cleaner. All overburden was removed so that only clear blue water remained above bedrock. Since the sheet piles around the cofferdam were all driven to refusal — could be driven no further — their top is irregular, paralleling the irregular contour of the bedrock. (Pier 17)

18. The aggregates or crushed stone for the Mackinac Bridge foundations had to be of specific sizes, hardness and cleanliness (actually washed two or three times to remove fine dust particles). It was quarried, crushed, screened and washed at Drummond Island, about 50 water miles away.

19. It was shipped in 12,000-ton self-unloading boats to the bridge site, where it was unloaded into the coffer-dams in three to four hours.

20. Prior to dumping aggregates into the cofferdam, vertical pipes were installed from the surface to the bottom. These pipes as shown were spaced about 20 feet apart and were known as grout pipes.

21. When a layer of aggregates about 10 feet deep was dumped inside the cofferdam, the four-story high marine grout plant was brought alongside. The plant made grout from cement, sand, fly ash, water and a patented intrusion-aid which gave it a smooth flowability and a pea soup consistency. This mixture was then pumped down the grout pipes to the bottom of the coffer-dam where the aggregates has previously been dumped. About 40 per cent of the area occupied by the aggregates was filled with water. Into this area the grout was pumped displacing the water, completely surrounding the crushed stone and hardening it into solid concrete.

22. Before the grout solidified, the grout pipes were pulled up to about two feet below surface of the crushed stone. Then more aggregates were dumped into the cofferdam, more grout was pumped in and the operation was repeated until the foundation reached the surface as shown.

23. A worm's-eye view of the top of the foundation after the two-inch steel dowels for connecting up with the superstructure had been installed. This surface was eventually concreted smooth and buried under 45,000 tons of concrete and steel in which the cables were anchored.

24. An aerial view of the south anchorage foundation, Pier 17, as it appeared in October, 1954, 85,000 tons of concrete and steel from the rock bottom of the Straits 90 feet below the surface to 10 feet above. To the right, foundation of Pier 18, caisson construction.

25. Snow, ice and sun combine with Pier 17 to make an artistic winter scene at the Straits of Mackinac.

26. By January, 1955, winter had completely shut down all work on Pier 17
and all other foundations as well. Viewed from above, the ice moving through
the Straits makes a design rivalling modern art.

While the anchorages known as Piers 17 and 22 were under con-struction, Merritt-Chapman and Scott was also building two huge foundations on which the main towers would be built. Also on the 1954 schedule were two smaller cable-bent piers, one a caisson, the other a pair of cofferdams. Here's how the main tower foundations, Piers 19 and 20, were toed into rock more than 200 feet below water surface, more than two miles from either shore, more enduring than the pyramids. [Pictures 27 through 40]

27. Loaded on a barge, here is one of the four parts of a corral—not to hold horses, but to hold a huge double-walled circular steel can. The structure, shown lying on its side, is made up of 20-inch pipes held together with six-inch pipe bracing. (Pier 19)

28. Three of the four caisson corral units were lowered to overburden. Through their 20-inch pipes heavy piles were driven to rock. Concrete was pumped into the pipes for additional strength. Next, the caisson was towed into position flush against the three units of the corral already anchored. Then the fourth side was installed. Thus the huge caisson was held in position while being lowered to the bottom, despite wind and high water. All movements were directed by surveyors from triangulation towers.

29. With the caisson guide (one side of the corral) resting on the bottom, workmen unhitch the block and tackle connected to the derricks which lowered this huge clumsy framework precisely into position. Note that rainfall does not deter the foundation builders.

30. A closeup of two of the four corral units showing 20-inch pipes with spud piles inside. These were driven to rock to anchor the framework sufficiently to contain the caisson.

31. The two caissons (French, meaning box, usually for artillery) were assembled on land at Rockport, Michigan, 90 miles from the Straits. Shown here is the incompleted bottom section. Note how the inside wall tapers to the outer wall to form a tough steel cutting edge.

32. Inside the caisson, welders sealed all parts so that it became watertight between the walls. These were then divided into eight water-tight compartments. (Pier 20)

33. When the caisson was built to a height of 44 feet (116 feet in diameter) a channel was dug from Lake Huron to the caisson, flooding the area beneath it. Being buoyant, the caisson could then be towed into Lake Huron and 90 miles northwest to the Straits of Mackinac.

34. Once in position the process of sinking the caisson began. This was accomplished by dumping rock aggregates into the space between the two circular steel walls. Acting as ballast, the stone caused the caisson to sink, and another double-walled ring of steel was added to the top as shown. (Pier 19)

35. Soon the bottom of the caisson reached the overburden. To penetrate this area "whirlies" with clam-shell buckets dug the overburden out of the center (the digging well) of the caisson. This reduced resistance against the cutting edges as the tapering wall at the bottom of the caisson weighted with added stone forced the overburden toward the center of the digging well—just like a giant cookie-cutter. The overburden was dumped outside the caisson.

36. Inside the caisson, the bottom of which was about 50 feet from bedrock, workmen welded the double-walled ring of steel. "Ring 16-19" identifies the sixteenth high stage of steel for Pier 19. Pier 20 is in the background.

37. Outside, the caisson men on three stories of scaffolding raced against winter to complete the driving of the caisson to bedrock. This photograph was taken on November 26, 1954.

38. The caisson bottom was about six feet from bedrock four days before Christmas, 1954. The driving was extremely difficult then, with penetration averaging only a few inches per day. Steam as well as explosives were required to break away boulders and other obstructing materials nearly 200 feet below the surface. (Pier 19)

39. The caisson was nearly founded in early January, 1955, as ice began to form in the Straits. Actually, winter was late this year and work did not shut down until January 14, 1955.

40. Crinkled ice makes an attractive pattern around the caisson still protected by the four corral units and the work platforms connecting them. Wind has piled ice up against one side of caisson while the opposite side is clear (upper right of picture).

The foundations for the two main towers were built with caissons. So was the foundation for the south cable-bent, Pier 18. However, this caisson was rectangular rather than circular, as are Piers 19 and 20, and instead of one center digging well, there were several, in fact, three rows of seven digging wells, each nine feet in diameter, thereby providing increased control during sinking. [Pictures 41 through 45]

41. The Pier 18 caisson, 44 x 92 feet, was assembled at Toledo to a height of 38 feet. It was towed to the Straits, positioned in its corral, and the process of driving it to rock bottom was the same as for a circular caisson. Here another lift of digging wells and wall is being welded to the top of the caisson.

42. The function of the corral is shown clearly in this picture of Pier 18 caisson. In view are two of the four units which anchor the caisson in place against wind, current and boat bumping.

43. The cutting edge of the caisson is nearly 100 feet below the surface and penetrating through overburden as clamshells dig it out from under through the 21 digging wells.

44. Driving the caisson for Pier 18 to rock gave the engineers and contractors a real scare when one of the inspectors discovered on a frostbitten midnight that the caisson was tilting. Piles were quickly driven to rock against the tilted side to prevent further distortion. Then the process of righting the caisson was undertaken by concentrating the loading and digging on the high side of the caisson. Here an engineer measures the angle of skew in support piles.

45. With caisson for Pier 18 founded in rock, it was solidly concreted the same as were the anchorage foundations, except that the concrete was discontinued eight feet below the surface. Two pedestals, 16 feet in diameter, with reinforcing rods sticking out, were built on top of the foundation, which is still in its corral as winter shows signs of closing in.

The sixth foundation undertaken in 1954 was for Pier 21, the north cable-bent tower. This foundation, comprised of a pair of circular cofferdams, each 50 feet in diameter, was a comparatively simple job, because it was founded on rock only 65 feet below the surface. [Pictures 46 through 52]

46. The steel framework for one of the pair of cofferdams took shape on land at St. Ignace. One of the units for a caisson corral is in the background.

47. The framework, swung aboard a barge, is transported to proper position in the Straits. It is 75 feet high and 50 feet in diameter.

48. When the twin steel frames were in position, sheet piling was driven to rock all around their outside circumferences. Then followed the same procedure for filling them with concrete as was applied in concreting the anchorages.

49. When the foundations were concreted to a height of eight feet below the surface, they were mounted with pedestals 16 feet in diameter. These in turn were filled with reinforcing rods.

50. The sheet piling was cut off eight feet below the surface. Here the real foundations are clearly revealed under water.

51. Winter wrote "finis" to construction operations for the first season, but the contractors did not give up without a struggle. Men chopping ice away to get out on the job November 17, 1954. They did not quit operations until January 14, 1955.

52. Late in 1954 looking north over the Straits showing the six important foundations undertaken in 1954. More important, they dramatize the physical line of the bridge, which up to that time had been largely a dream. Note the causeway jutting into the Straits, upper right, built in 1941.

III
Sky-bound Steel Towers
and Rock-bound Foundations
1955

Throughout 1954 shorewalk superintendents wondered whether or not a bridge was really being built at the Straits. Certainly, there were lots of boats and complicated equipment huffing and puffing between St. Ignace and Mackinaw City, but nothing resembling a bridge appeared to be going up in the Straits. This was primarily because everything was going down—down to bedrock. Of the 1,000,000 tons of concrete and steel in the Mackinac Bridge, 750,000 of them are under water. However, 1955 was the year of the sky-bound steel towers and the 10-story high superstructures. [Pictures 53 through 63]

53. As soon as the ice was sufficiently broken up so that marine equipment could navigate, Merritt-Chapman and Scott built a 50-foot high trestle on top of each anchor foundation (not visible) and then raised a crane to the top of the trestle to facilitate the handling of materials.

54. Next came the erecting of forms to hold the concrete that would hold the steel that would hold the cables that would hold the suspension bridge.

55. An overall view showing rising center walls of anchor block superstructure and crane trestle about half buried in concrete. The story goes that the first crane operator assigned to this job quit as soon as he saw the perch on which he had to work. (Pier 17)

56. This three-story high chunk of steel is called an "Anchor Bar Support Frame," and that is just what it does: supports the anchor bars (see next picture). (Pier 22)

57. The American Bridge Division of the U. S. Steel Corporation erecting the first steel in their $44,532,900 contract for the Mackinac Bridge superstructure. This first steel, consisting of anchor bars, is all buried in concrete.

58. All the anchor bars are in place. To them will be attached additional links around which the individual wires making up the cables will be looped. Note the height of the center wall almost up to the top of the trestle. (Pier 22)

59. Another view of the anchor block and the eye-bars dramatizes their function as they are huddled together forming a strong compact group designed to resist the 30,000-ton pull of the cables. (Pier 17)

60. As the 1955 summer wore on the superstructure continued upward, burying the anchor bars and the trestle and gradually taking on some appearance of its function. (Pier 22)

61. When winter approached, the anchor bars were solidly concreted in place. The boom of the crane is barely visible as the walls around it nearly reach their full height of 118 feet. (Piers 17 and 18)

62. These were days to try the durability of the strongest men and equipment as well. The men usually proved the more durable. Here, a piece of tough equipment lies, wounded, on its side as it tried to buck a gust of wind with too large a load.

63. Another manhandled piece of equipment that buckled under a misplaced load. The ancient ore carrier *Wolverine* was used to store sand. Too much got into too little an area and the fatigued steel quit. Believe it or not, after the sand was unloaded she straightened out and with minor repairs was back in floating service again.

As with the anchor blocks work resumed on the main tower foundations as soon as the ice melted. [Pictures 64 through 82]

64. Notwithstanding the fears of the faint-hearted who were certain the bridge work would never stand the Straits ice, there was no ice damage whatsoever. Here is one of the incompleted caisson foundations. A twin-walled steel can without concrete reinforcing challenged the winter's worst. And when spring came, inspection revealed that these caissons came through unscathed. (Pier 19)

65. The month of May, 1955, is one not easily forgotten. Closed out by the winter before, the caissons for the main tower foundations were left unconcreted and somewhat behind schedule. When work resumed in the spring of 1955, the concreting began, and during the 31 days of a pleasant, mild, dry May 103,000 cubic yards of concrete were poured into Mackinac Bridge foundations, setting a new record for underwater consolidation of concrete. (Pier 20)

66. Notwithstanding its hodge-podge appearance, this is really an orderly construction scene. Around the outside of the center are the four corral units still being used as work platforms. They no longer hold the caisson in place. Not only is the caisson founded on rock, but it is filled with concrete. Inside the center circle are two smaller circles, 38 feet in diameter. They are the pedestals that will arise from the foundation eight feet below the surface. (Pier 19)

67. The two pedestals on top of the foundation are 25 feet above the water surface. The bolts embedded 20 feet in these concrete pedestals were sticking up ready to grab hold of the steel for the towers.

68. Meanwhile the towers were not only being fabricated but were being built—horizontally instead of vertically—some 500 or more miles from the Straits of Mackinac. At Ambridge, Pennsylvania, the towers were laid out on the ground and drift-pinned together to see if all parts fitted properly. This picture showing the bottom section provides a clue to their enormous size.

69. Finally came the day when actual tower erection could begin. The first piece was hardly impressive, a base plate just four inches high, though it weighed 13 tons.

70. Next came the actual tower section weighing 40 tons and representing the bottom 16 feet of a tower that rises 552 feet above the surface.

71. Engineers, pushers and journeymen steel and ironworkers prod and push until the giant sections fit properly over the bolts which will secure it and 6,500 additional tons of tower steel to the pedestal, which in turn is connected to the foundation that goes down more than 200 feet to rock.

72. It was exactly 6:10 and 15 seconds in the evening of July 2, 1955 when the first piece of vertical tower steel slid into place. Obviously, all are pleased as the project manager congratulates the crew foreman.

73. A derrick mounted on a barge erected additional tiers of tower steel. Work was slow because wind, waves, wakes and current rocked the barge and load as well. Consequently, placing heavy steel sections precisely in position required consummate skill and lots of time.

74. As soon as three tiers of steel on each tower leg were erected, a creeper was attached. This ingenious device mounted with a boom raised steel and climbed up the steel it raised. No wind, water, wakes or current to worry about —just height.

75. Three and a half months later the tower steel reached the 420 foot mark. Sections were bolted together temporarily as riveters patiently worked away on the 6,000,000 rivets designed to hold the towers together. Note riveters' working cages at the height of the first strut.

76. A close-up of the towers reveals their unlimited possibilities for photographic angles. Note the wire cage in the lower center of the picture. This is a personnel elevator. It wasn't bad except on real windy days when it was occasionally blown out of its guide lines.

77. Inside the tower were the men who backed the rivets driven from the outside. Their only light came from flashlights and the miner's lamps attached to their safety helmets.

78. Finally, on October 22, 1955 the first topmost piece of steel was erected, and according to tradition Old Glory waved proudly with the men who put it there, 552 feet above the Straits of Mackinac.

79. The Straits could be rough even during the construction. Here waves and spray more than 40 feet high pound fruitlessly against the Mackinac Bridge towers and foundations.

80. Safety was first, last and always with Mackinac Bridge contractors. Consequently, a splendid record of comparatively few casualties and fatalities was made during construction. However, here is what happened when a rivet fell off the tower onto the deck of one of the boats.

81. Both towers completely erected (not yet fully riveted), mounted on top with 80-foot high booms to handle materials, were a pleasing sight on a calm day.

82. Bridge builders with lunch in hand, electric cap lights lit, and life preservers secure pull away from the dock to start another day's work.

While the extremely interesting spectacle of tower construction was carried on during 1955, there was other work in progress of a less striking nature. The superstructure for the anchorages housing the massive anchor bars has already been pictured. In addition, the superstructure for the cable-bent piers was erected. [Pictures 83 through 88] Twenty-seven additional foundations and their superstructures were built, and just as winter rang down the ice curtain on the 1955 construction season two huge side spans, the first horizontal steel, were floated into position.

83. The south cable-bent foundation Pier 18, caisson construction founded on rock 130 feet below the surface, was ready for its superstructure early in 1955. A piece of a corral unit was placed on the foundation between pedestals as a work platform.

84. Circular forms 16 feet in diameter were filled with concrete at regular stages to build twin towers that look like a couple of marine silos.

85. A massive steel arch was raised to the top of the concrete towers and made a bridge connecting them.

86. Forms were built around the arch frame, and they were filled with concrete.

87. Steel towers, over which cables did eventually bend as they swept down from towers toward the anchorages, were erected on top of the concrete superstructure.

88. The superstructure construction on the south cable-bent was duplicated on the north cable-bent Pier 21. Only their foundations were different. In the picture the derrick was preparing to lift an empty form off the barge onto the pedestal at the extreme right.

The 27 remaining marine foundations were cofferdam construction, the same as the anchorage piers or the twin cofferdams for Pier 21. However, five foundations were built in an area where rock was comparatively deep for the size of the foundation required. Efficient construction methods dictated that the cofferdam be built on steel spud piles driven through the overburden to rock. Piles had to be driven at angles to provide a wide supporting base. [Pictures 89 through 95]

89. The steel framework 90 feet high for one of the five cofferdams on spud piles was erected on land. (Pier 14)

90. It was obviously a beautiful calm day, that May 13, 1955, when the framework was towed past the former ferry docks to its position in the Straits.

91. Once in position, sheet piling was driven all around the framework, but not to bedrock or refusal, because a different method was being used on this foundation.

92. Steel H-beam piles, 96 of them, were driven inside the cofferdam to rock and refusal. Many of them were battered (driven on angle as shown here) to provide a wider and thereby stronger base to hold the concrete load to be placed therein. (Pier 11)

93. A single shaft of concrete emerged from the 50-foot diameter foundation lurking eight feet below water. A hammerhead frame (sometimes called elephant ears) was placed on top of the shaft. Forms were built around the frame and it was completely encased in concrete. (Pier 26)

94. South from the south anchorage, 16 foundations come into view,
one exactly behind the other as straight as an arrow. (Piers 1-16)

95. Also south from the end of the causeway at St. Ignace, the north foundations, 11 of them, line up to form stepping stones across the Straits.

Rivalling the towers for bold and spectacular construction was the placing of the side spans. These are the horizontal steel stiffening trusses designed to support the road deck between the anchor blocks and the cable-bents. They are 472 feet long, weigh 720 tons, and were mounted on falsework 80 feet high, which in turn was mounted on two barges lashed together. A real top-heavy arrangement that could be towed into position only in calm weather. [Pictures 96 through 102]

96. November 10, 1955, the south backstay span was just about ready for launching, but the weather wasn't. On the sixteenth, a blizzard with winds gusting up to 76 miles per hour nearly toppled the span.

97. On the nineteenth, the Straits were calm, and with thumbs crossed hoping and praying the weather would remain calm, the contractors ordered "operation backstay span."

98. With tugs pulling and pushing and winches wheezing with taught wire ropes, the backstay was slowly babied into position.

99. A close-up shows there were about six feet to go before the back-stay span lined up with the steel shoe into which it would eventually fit.

100. The holes in the span were flush with the holes in the shoe and a 500-pound pin was rammed into these holes to secure the backstay to the anchorage.

101. Pinned to the anchorage, left, and sitting firmly on steel shoes on the cable-bent, right, the backstay span was made clear of the falsework when the barges on which it was built were flooded and consequently lowered, so that they, along with the falsework, could be towed out from under the span.

102. Exactly the same operation was performed on December 18, 1955 to place the north back stay span, shown here with the north anchor block as winter weather moved in. (Pier 22)

IV
Cable Spinning, Truss Span Erecting and Deck Paving
1956

Spinning the cables aroused more curiosity and stimulated more questions than any other phase of Mackinac Bridge construction. It was the highlight of 1956 operations. It was a chancey operation in that once it was started, it had to be completed. Partly finished cables could not be left open to the rigorous Straits winter. Consequently, when cables were completed in the record time of 78 working days, all concerned breathed and slept more peacefully. Other construction in 1956 consisted of the erection of truss spans out from each shore and the partial paving of roadways on these spans. [Pictures 103 through 138]

103. In order to spin the cable there had to be a catwalk on which the cable spinners could work. In order to build a catwalk there had to be wire ropes to support it, five for each catwalk, raised from one anchor block over the tower tops to the other anchor block following the same curve and contour that the cables would eventually take. Here the wire ropes were started from the anchorage to the cable-bent. (Piers 21 and 22)

104. Wire rope was unreeled from a barge towed across the Straits. Rope was two and one-quarter inches in diameter. After serving as catwalk support it was re-employed later to hold up the suspended spans.

105. The ropes were pulled over the tops of the tower. Note the huge steel frameworks and platforms built on top of the tower to take care of the equipment and operations necessary for cable spinning.

106. With the wire ropes in place, the next job was to slide the cyclone wire fence, which was really the catwalk, down the wire rope from the tops of the towers toward the cable-bents and toward the center of the main span.

107. Slowly but surely accordion wrapped sections of cyclone fence unfolded and slid gently down the wire ropes.

108. Coming down from each tower in opposite directions toward the center of the main span, the catwalks were nearly joined together.

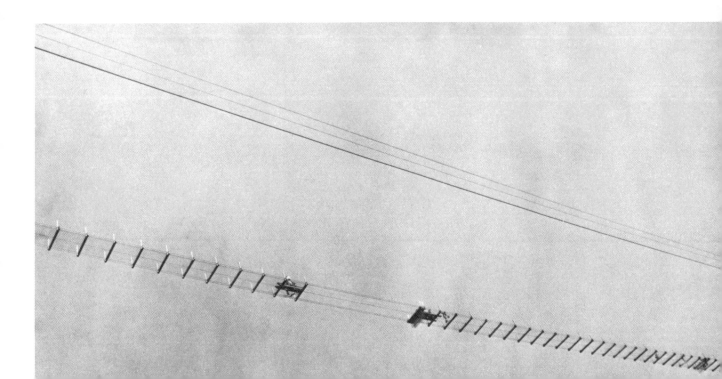

109. Bridge workers were pleased to have their picture taken as they joined hands in the center of the main span 190 feet above water, signifying that the cyclone fence catwalk had been connected from anchorage to anchorage.

110. Looking through the 13-ton cast iron saddle in which the 12,580 cable wires will be held, one can see the completed catwalk ahead, and the incomplete one to right of center.

111. Bridgemen had to do all the hand work and detailed work that went into making the catwalk a safe and useful platform on which to work.

112. Wood cleats were wired to the fence floor to secure safer walking on the catwalk. Handrails were erected and steel trusses connected the catwalks for additional steadiness and sturdiness.

113. Storm guys were attached to the catwalk securing it to the towers and the anchor blocks to help resist high winds and storms. The billowing sails of the yacht below complement the curves of the catwalk and the storm guys.

114. A close-up of some of the machinery mounted atop the towers to take care of the cable spinning.

115. Finally the catwalk was completed and ready for the cable spinning. Some elderly lady wrote to the Authority and demanded to know how the members expected a car to climb the high, steep road they had built across the Straits.

116. Beginning in 1955, 55,500 coils of steel wire for the cables had been shipped for storage to Sault Ste. Marie, Michigan. Buildings were specially erected to store these coils prior to their being rolled onto reels.

117. In the reeling building, the coils were spliced and wound automatically under precisely measured tension on to huge drums.

118. Each reel weighed 16 tons and contained approximately 320,000 feet of wire. The wire was much stiffer than its appearance would indicate and was .192 inches in diameter—about as thick as a pencil.

119. Logistics required that a large stockpile of wire already on reels be ready before cable spinning began. These reels were shipped by rail to the Straits from the Soo. Empty reels had to be rewound as soon as they were returned so that the cable spinning would not be delayed.

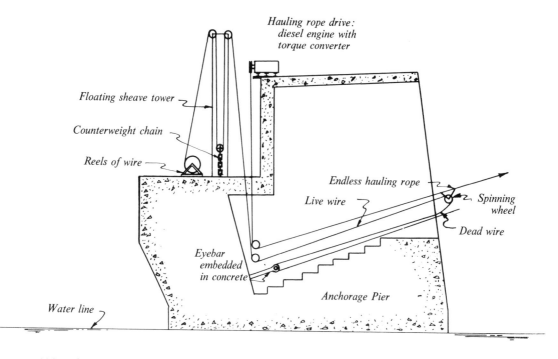

Hauling rope drive:
diesel engine with
torque converter

Floating sheave tower

Counterweight chain

Reels of wire

Endless hauling rope

Live wire

Spinning wheel

Dead wire

Eyebar
embedded
in concrete

Anchorage Pier

Water line

120. Diagram shows the key parts of cable spinning, **which is** a very simple process and sadly mis-named, since there is no spinning, braiding or twisting involved, as most **people** suspect.

121. Derrick raises 16 tons of cable wire onto huge anchorage spinning platform as though the reel were a spool of thread.

122. A bridgeman walks the double-grooved spinning wheel from anchorage to anchorage to ascertain that it is properly working before the machinery is pressed into action.

123. The first four wires are pulled out from the north anchor block on their 12-minute trip up and down the two towers to the south anchor block.

124. Moving down from the top of the tower, the spinning wheel laid down four wires to be included in a strand of 340 wires. Thirty-seven strands make up the 24¼-inch cable.

125. As soon as the spinning wheel arrived at the anchor block the four wires it carried from the opposite side were slipped around strand shoes (lower left and center) and four wires from the reel on this anchor block were looped over the wheel and it was sent on its way.

126. Since there were two spinning wheels operating simultaneously in opposite directions, they would meet exactly in the middle. When their trips were completed eight wires had been added to the cables.

127. Each and every wire pulled across the suspension span was individually adjusted. Strands of wires were adjusted usually at night when the temperature was constant and the wires not likely to contract or expand depending on the location of the sun.

128. Usually five strands of wire were spun on each cable at a time. While the spinning was taking place on one cable, bridgemen adjusted, squeezed and banded the strands on the other.

129. Cable spinning was carried on in two ten-hour shifts so long as weather permitted. The lights erected for the night shift proved so attractive that the Bridge Authority made them permanent fixtures of the bridge.

130. With the cable spinning only a few days from completion, the photographer caught one of the Pittsburgh Steamship Division fleet, U.S. Steel, saluting the cable spinners. Much speculation arose as to whether or not this ship had carried any of the ore for making the steel out of which the cable wire was made.

131. When the cable spinning was completed the chains to which the strand shoes were attached were embedded in concrete .(Pier 22)

132. Thirty-seven strands of 340 wires each were compressed into a circular shape and bound with stainless steel straps to form the more-than-two-feet-in-diameter cable.

133. Engineers carefully measured the diameter of the cable to make sure it was properly compacted. Thus every steel wire would do its equal share of work in holding up the suspended spans.

134. As soon as the cable was compressed and banded the catwalk was attached to it. The wire ropes which originally supported the catwalk were removed and sent to Trenton, N. J., for stretching, cutting and cupping into the precise lengths for use to hold up the suspended span.

135. Cable bands or clamps which hold the cable compact and cradle the suspender ropes which loop over them, connecting the suspended bridge with the cable. The center cable bands are in the foreground.

136. Bridgemen work the ratchet that pulls the wrench that tightens the center cable clamp. They are working on the catwalk 190 feet above water.

137. November 10, 1956 and bridgemen fight freezing temperatures as well as snow and ice to tighten cable clamps 500 feet above the Straits.

138. Finally winter closed in, as it always does, but the cables were securely in place and another key operation in the construction of the Mackinac Bridge was completed on schedule.

While the cables were being stretched across the suspension bridge, the not inconsiderable truss span bridges were being erected. These twenty-seven spans, some almost 600 feet in length, would have been considered sizable bridges anywhere. In the Straits, with the constant winds and unpredictable storms, they represented a noteworthy achievement, since they were completed without incident. It was only because the cable spinning was considered dramatic and romantic that it diverted attention from truss span erection. [Pictures 139 through 148]

139. Northward from Mackinaw City, steel truss erection moved Straitsward from Pier 1 to Pier 2. Date, April 24, 1956.

140. Two and a half months later the truss spans were erected out to Pier 7, some 1,443 feet north of the starting point.

141. Construction began from the end of the causeway on the north side of the Straits in southerly direction in mid-May of 1956.

142. Falsework had to be erected between distant piers to support the truss spans until all the steel was "hooked up," connecting one permanent pier to the next.

143. While machinery and labor saving devices do the heavy work, bridge erection is largely a handmade job. The final movements of every connection are made by hand and all drift pins as well as steel bolts must be individually inserted and tightened. (Pier 24)

144. Bridgemen tightening high tension steel bolts used for all truss span connections. Meter in torque wrench must be carefully observed to make sure that a bolt is neither too tight nor too loose. Every bolt was inspected.

145. A close-up of truss work construction shows some of the detailed operations. Frequently shorewalk superintendents could not see anybody working on the bridge. Can you spot the eleven men in this picture?

146. By October 1956 the truss span erection had reached out to Pier 12 some 3,749 feet north of the Mackinaw City shore. Note the road deck getting underway.

147. Some days were rougher than others and one morning when the contractors came to work they found, among other things, that their big goose neck crane had ripped its moorings in a Straits storm and headed for parts unknown. Fortunately, it beached on Mackinac Island only a few miles away and sustained no damage, except perhaps to the sign on high and the pride therein.

148. Contractors were confident that they would erect the steel to connect Michigan's two peninsulas before the end of 1956, but weather had a different view and prevailed, as weather often does. Work had to be stopped in late December just 325 feet short of the goal on both sides of the Straits. (Note gaps at anchor blocks.)

The construction schedule called for paving the road deck from the shore half way to the anchorages in 1956. By comparison with other construction challenges on the Mackinac Bridge, paving the deck was a simple operation, under difficult and crowded conditions. However, when compared to paving a highway at ground level, it was a substantial challenge. This was evidenced by the fact that contractors did not flock to bid on the job. [Pictures 149 through 155]

149. Deckwork got underway on the north side in early August 1956. Six inches of concrete, plenty of reinforcing steel, topped with two inches of asphaltic concrete, commonly known as blacktop.

150. Pavers leveled off concrete as a foreman was about to signal the man operating the Georgia Buggy to dump another load. Note the intent expression on all the eleven men as they team up to do a job.

151. (Above right) About 1800 feet of four lane deck was completed on the north side before winter called a halt to operations in 1956. Picture was taken looking south from the end of the causeway.

152. Whereas concrete was conveyed to the pavers with trucks or Georgia Buggies on the north side, the south side contractor used a pumpcrete machine. This could pump concrete through a pipe about 1100 feet, thus eliminating wheeled conveyances. Note the reinforcing steel mesh and plywood forms for concrete.

153. Concrete pours out of the pumpcrete pipe to fill the forms for paving the deck on the south side. Note the expansion dam in right foreground.

154. On the north side the connection from the bridge deck to the causeway level was started with the completion of 14 bents on which steel and roadway would be built.

155. On the south side the connection between the bridge deck at Pier 1 south to the new interstate highway was accomplished by 14 bents on which steel was erected.

V

A Million
Final Things
1957

It seemed impossible that the bridge could be completed sufficiently by November 1, 1957, to carry traffic. There were still a million things to do. Looming largest of all was the raising of the suspended span, bolting it together, and paving it. Along with this major project the decks had to be paved out to the anchor blocks. The anchor blocks had to be filled with concrete to the full 118 feet of height, the cables wrapped with steel wire, the lighting installed, and a score or more of minor problems solved, such as the building of the Administration Building, the toll plaza, the toll collection booths and the paving of the causeway from the toll plaza to Pier 34, which is the northernmost foundation.

156. The wire rope which in 1956 had been used to support the catwalk had been shipped back from Trenton in carefully cut lengths wound on reels. Each individual wire rope was marked for a precise position on the cable from where it would hang to hold a portion of the suspended span.

157. The job of raising these wire ropes was begun by positioning a barge directly under the cable band over which the wire rope would loop and then pulling it up off the barge to the cable band.

158. A bridge worker nudges the steel cup in which the end of the suspender rope is buried on its way down so that the rope will hang in equal lengths from the cable where it has been looped around the cable band.

159. Bridgeman slides down suspender to tighten the clamp during installation of the two and one-quarter wire rope. He is securely tied to the catwalk, about 500 feet above the water.

160. With nearly all the suspender ropes hanging from the cable the suspended spans adjacent to the towers were cantilevered out and attached to the suspender ropes. The cups at the ends of the suspenders were used for attaching the suspenders to the suspended span.

161. Meanwhile, back at the St. Ignace dock, the superstructure contractors were assembling portions of the suspended span. This work went on all during the winter of 1956-57.

162. The sections, about 120 feet long and 40 feet high, were assembled on tracks so that they could be rolled down toward the shore where barges would be ready to receive them as soon as the ice went out of the Straits.

163. The first of these suspended portions was towed into position on June 5, 1957. Block and tackle were attached at four points and upon signal from the superintendent the delicate process of raising the section was begun.

164. Many an experienced bridge builder watched with bated breath as the slow, steady, co-ordinated process of raising the suspended span portion, or stiffening truss, was carried out.

165. The stiffening truss sections had to be raised on a carefully calculated schedule so that the weight on the cables did not get out of balance. Because the suspended portion of the bridge did not take its final shape until all the weight had been placed on it, the contour of the partially erected suspension bridge seems completely distorted.

166. The signal man "talks" the stiffening truss into position as bridgemen wait on it ready to insert the drift pins as soon as the key holes between units are matched.

167. A closeup of the lifting strut, which was moved down the cables and secured in position over each cable band in order to lift each stiffening truss section straight up to where the suspender rope was waiting to receive it.

168. Finally, by July 22, 1957, all but three of the stiffening truss sections had been raised: two at each cable-bent and one huge section in the center. Note the curve of the suspended bridge as distinguished from the curve of the partially suspended bridge.

169. The breeze was brisk and the water rough despite the presence of some 75 news correspondents who came to witness the raising of the center section. Due to unsympathetic weather it was not accomplished until nearly dark.

170. Just about 40 more feet to go and bridgemen will bolt the center section of the stiffening truss to the already erected portions extending toward the middle, and Mackinac Bridge steel will be connected.

171. Notwithstanding the fact that these men are coming off the job after a ten-hour day, they express the rugged determination and capability necessary to complete a job the size of Mackinac. All of them were proud to have had a hand in it.

172. As soon as the vertical stiffening truss was completely erected bridgemen laid in about 80 feet of deck on each side of each tower. Using this as a work platform a 40-ton crane was raised as though it were a child's toy.

173. Work progressed rather rapidly on the construction of the steel bridge deck. It had been designed for rapid installation and the design was paying off in August as the scheduled opening date was only three months away.

174. Open grid grating was used for the inside center lanes of the bridge, and steel forms for holding light-weight concrete were installed for the outside lanes. The center of the picture shows a Ross Carrier transporting five huge stringers over newly installed deck to the place where they will be strung out on the stiffening trusses to support additional roadway deck.

175. A crane lifts a roadway stringer into position as other bridgemen install curbs, railing, and put finishing touches on roadway grating.

176. By early September 1957 the road on the suspended bridge extended nearly all the way from each tower to the center.

177. On September 9 the last section of the center grid deck at the north end of the suspended span was lowered into position.

178. On September 10 the last section of the center deck grid at the south end of the suspended span was lowered into position. Now there were less than two months left for paving, cable wrapping, miscellaneous painting and cleanup on the suspended portion of the bridge alone.

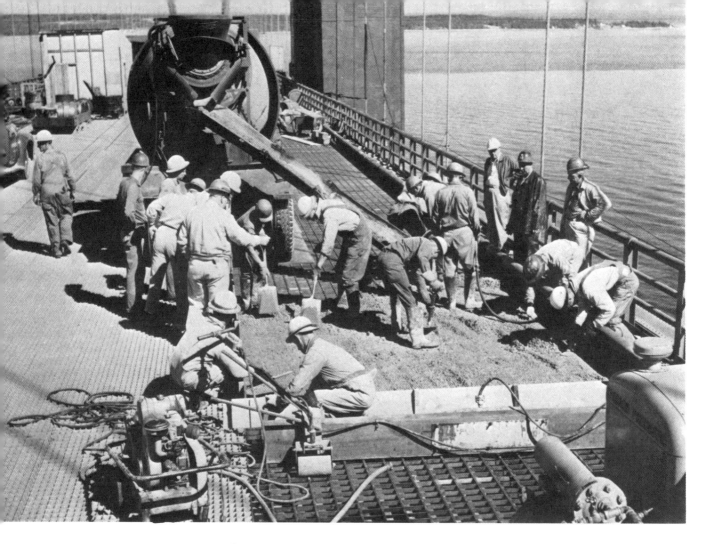

179. Paving of the suspended span roadway began in mid-September as lightweight concrete was poured into the steel forms of the outer lanes.

180. All during October bridgemen welded more than a million connections to permanently attach the open grating portion of the roadway to the floor beams on which they rested.

181. Actually it wasn't until a day or two before the bridge opening that the blacktop paver finally finished laying asphaltic concrete over the lightweight concrete on the outside lanes of the suspension span.

182. While all the work on the roadway was going on, men were strung out overhead on eight sections of the cable applying red lead paste and wrapping pencil-thin steel wire as tightly as possible around the cable to provide a tough watertight cover against the elements.

183. A closeup of cable wrapping operation shows an intricate machine on which three bobbins feed the steel wire that covers the cable.

While work progressed steadily on the suspended span the job of pouring 25,000 yards of concrete into the anchorage superstructure had also to be completed in 1957. However, in order to accomplish this, the remaining 325 feet of steel truss span on either side had to be erected and the concrete roadways brought all the way out to the anchor blocks.

184. At the outset of the 1957 construction season the previous year's unfinished work was completed when the truss spans were carried all the way to the anchor blocks, thus completing a steel structure across the Straits.

185. On May 17, 1957, a steel chord was lowered from the truss span to the rear wall of the south anchor block and for the first time there was a continuous steel bridge across the Straits. (Pier 17)

186. Although these steel chords, which marked the last piece of the first steel to span the Straits, were only about 22 inches wide they apparently provided a safe and secure footing for experienced bridgemen.

187. Deck contractors pulled out all stops to complete the concrete roadway all the way out to the cable-bents where the steel and lightweight concrete roadway begins.

188. The concrete deck on the approach spans was covered with two inches of black-top, which operation was carried on during the last few days of October.

189. There was still one more huge and critical job to be done, and that involved the completion of the anchor block. Solid concrete had to be poured into the rear third of this block while the walls around the cable splay areas had to be raised to deck level. Here on July 10 Merritt-Chapman and Scott built two foot bridges from the approach span roadway to the completed wall of the anchor block, 118 feet high. (Pier 22)

190. By early August the rear portion of the anchor blocks had been raised to their full height and the walling-in process had begun.

191. An interesting and unusual device, known as an "elephant trunk" was used to direct concrete from the top of the block to the area some 60 feet below, where it was dumped.

192. Here the cables splayed out for connection with the strand shoes. They were covered with huge planks against damage from fire and falling articles. In late August the walls around this splay area began to rise.

193. The deck of the anchor block was made up of 13-ton slabs of reinforced, pre-cast concrete. These were poured during the summer but were not placed until the walls were completed in late October.

194. By the end of October the cable splay area chamber was complete. The cable enters the anchor block at the right through the cable saddle, and strands of the cable are looped at the left around strand shoes embedded in concrete.

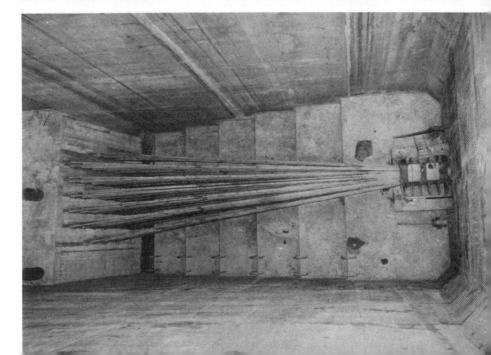

195. These American Bridge Division superintendents certainly displayed the satisfaction they must have felt in knowing that they had made the critical opening day deadline of November 1, 1957.

196. Also during the 1957 season the Authority's Administration Building was completed. The facing on the building is Drummond dolomite stone, the same as that used in the bridge foundations.

197. During the 1957 season the electrical contractor installed hundreds of miles of cable, 248 light standards, and more than 70 navigation lights as well as aerial beacons to complete the electrical requirements for the five-mile-long project.

198. The Mackinac Bridge Authority toll plaza is designed to accommodate nine toll booths and ten lanes of traffic. Five are installed at present to take care of six lanes of traffic. Under extreme conditions five lanes of traffic in a given direction can be moved through the toll booths.

199. The Michigan State Highway Department completed in 1957 work begun in 1956 to provide modern divided-lane highway approaches and connections from existing highways to the Mackinac Bridge.

200. Finally, on November 1, 1957, the bridge was opened to traffic. It was a happy day for all of Michigan, and expressions on the faces of these individuals largely responsible for the structure are witness thereof. They are, from left to right, John C. Mackie, Michigan State Highway Commissioner; Dr. D. B. Steinman, Consulting Engineer and designer of the Mackinac Bridge; Governor G. Mennen Williams; Prentiss M. Brown, Chairman of the Mackinac Bridge Authority; Murray D. Van-Wagoner, George A. Osborn, William J. Cochran, members of the Mackinac Bridge Authority; and Lawrence A. Rubin, the Authority's Executive Secretary.

201. Viewed from the ground, the towers of the bridge are not unlike the spires of Michigan's stately pines, while the truss spans may be compared to the green needle-decked boughs. The colors of the Mackinac Bridge are foliage green for the horizontal steel and ivory for the towers.

202. To symbolize the tying together of Michigan's two peninsulas by the Mackinac Bridge, Governor Williams and Authority Chairman Brown and wives tie two pieces of ribbon stretching 22,700 ft. from shore to shore at the dedication on June 28, 1957.

203. Brilliant displays of fireworks high-lighted the evening festivities during the four-day celebration marking the Bridge dedication.

204. Moonlight and Mackinac — in an area already renowned for its natural beauty Mackinac Bridge by night, as well as by day, adds another jewel in the diadem of nature's wondrous way.

APPENDIX

MACKINAC BRIDGE ENGINEERS AND CONTRACTORS

ENGINEERS

Consulting Engineer, David B. Steinman, New York City
Consultant, Glenn B. Woodruff, San Francisco

Harley, Ellington and Day, Detroit—*Architects for the Administration Building under the direction of Dr. Steinman*

CONTRACTORS

	Amount of Contract*
American Bridge Division United States Steel Corporation, Pittsburgh	$43,927,806
Entire steel superstructure, including towers, cables, suspended and truss spans, and viaduct in Mackinac City.	
Merritt-Chapman and Scott Corp., New York	$26,335,000
33 marine and one land foundations mounted with piers and borings at all foundations.	
Louis Garavaglia, Center Line, Michigan and Johnson-Greene Company, Ann Arbor, Mich.	$ 2,181,093
Paving of entire bridge structure.	
Blumenthal-Kahn Electric Co., Inc., Baltimore	$ 794,894
Installation of deck, cable, and navigation lights and all electrical work.	
Louis Garavaglia, Center Line, Michigan	$ 494,887
Paving of toll plaza and causeway.	
Omega Construction Co., Grand Rapids, Mich.	$ 232,500
Construction of Administration and Maintenance buildings.	
Taller and Cooper, Inc., Brooklyn	$ 149,624
Erection of toll booths and installation of toll collection and recording equipment.	
Acme Elevator Company, Detroit	$ 78,285
Installation of elevators in each of the towers.	
Jersey Testing Laboratory, Newark	$ 70,000
Testing of all steel used in bridge construction.	

	Amount of Contract*
Durocher and Van Antwerp, Cheboygan, Michigan	$ 60,600
Extension of causeway.	
Pittsburgh Testing Laboratory, Detroit	$ 25,500
Testing of cement used in bridge construction.	
Erkfitz Plumbing and Heating Company, Rogers City, Michigan	$ 22,418
Heating and mechanical installation in Maintenance Building.	
Edison Sault Electric Company, Sault Ste. Marie, Michigan	$ 19,000
Supplying extension of water supply to bridge site.	
Hatzel and Buehler, Inc., Lansing, Michigan	$ 11,000
Electrical installation in Maintenance Building.	
Fred Hoffman, Petoskey, Michigan	$ 7,000
Tree planting and landscaping.	
Straits Construction Co., St. Ignace, Michigan Alpine Construction Co., St. Ignace, Michigan Denison Construction Co., Munising, Michigan Thornton Construction Co., Houghton, Michigan and O. O. Snowden, Escanaba, Michigan	$ 33,343
Miscellaneous grading, sewer, curb, paving, blacktopping, and guard rail construction and installation.	
Miscellaneous small contracts and extra work are not included above.	

*Due to the fact that all contracts are not yet complete, the figures shown may vary from those certified by the engineer when the construction account is finally closed.

The Piers of Mackinac Bridge located and numbered

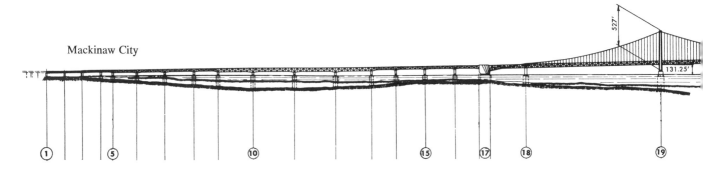

Mackinaw City

THE MACKINAC BRIDGE AUTHORITY

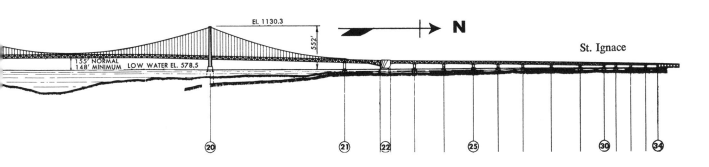

El. 1130.3

552'

N

St. Ignace

155' NORMAL
148' MINIMUM   LOW WATER EL. 578.5

20   21 22   25   30   34

MACKINAC BRIDGE AUTHORITY 1978

CHARLES T. FISHER III
Chairman

Born in Detroit, A.B. Georgetown University 1951, M.B.A. Harvard 1953. President and Director, National Detroit Corporation and National Bank of Detroit. Director, International Bank of Detroit, General Motors Corporation, American Airlines, Detroit Edison Company and Hiram Walker-Gooderham & Worts, Ltd. Member Michigan Association of Certified Public Accountants and American Institute of Certified Public Accountants. Trustee and Secretary Automotive Safety Foundation, Director and Secretary Highway Users Federation, and Trustee, Treasurer and Member Executive Committee American Enterprise Institute. Appointed to Mackinac Bridge Authority October, 1967.

RALPH B. K. PETERSON

Born in Chicago, Illinois; attended Augustana College in Rock Island, Illinois; graduated 1961 with Bachelor of Arts degree: attended Northwestern University Law School and was awarded a Juris Doctorate in 1964; was in private practice in Chicago until 1968; moved to Escanaba, Michigan, appointed as City Attorney for the City of Escanaba 1970; presently member of the Hansley, Neiman, Peterson, Beauchamp, Stupak & Bergman, P.C. law firm. Appointed to the Mackinac Bridge Authority on January 1, 1974.

GEORGE W. WILSON

Born Knightstown, Indiana; Bachelor of Arts Washington & Lee University, 1939; joined Peoples Bank and Trust, Alpena, Michigan 1957 as President, elected Chairman of the Board 1977; previously with United California Bank and Michigan State Banking Department; 1942-1946, First Lieutenant, U.S. Army; President, Michigan Bankers Association 1978, Member of Detroit SBA Advisory Council, Greater Michigan Foundation, Alpena Industrial Development Corporation, Board of Deacons First Congregational Church; Governing Council, American Bankers Association. Appointed to Mackinac Bridge Authority April, 1972.

WHEELER J. WITTE

Born in Iron Mountain, Michigan. World War II 893 Air Engineers 9th AAF; Waling delegate and Vice President of Timberworkers Union in Wisconsin and Michigan; Member of Society of Professionals in Dispute Resolution and Industrial Relations Research Association. Life member of Disabled American Veterans. Mediator, Bureau of Employment Relations, State Department of Labor. Former President of Ford Local No. 952, UAW CIO; President of Iron Mountain Vicinity Industrial Union Council CIO; Financial Secretary-Treasurer, Building Trades, Dickinson County, AF of L; Secretary, Grand Rapids Building Trades Council AFL-CIO. Served on Mackinac Bridge Authority from 1967 to 1970.

LOREN E. MONROE

Born in Thomasville, Georgia, graduated from Wayne State University—B.S.—Accounting 1958; J.D.—1970, admitted to Michigan Bar 1970; received CPA Certificate in 1974. Field Auditor, Michigan Department of Treasury from 1959 to 1970 and tax specialist at Coopers & Lybrand, International CPA firm, from 1970 to 1976. Appointed State Treasurer and Treasurer of the Authority in 1978.

ORIN K. GRETTENBERGER

Born in Ingham County, Michigan. B.S. Ferris Institute 1931, M.S. Philadelphia College of Pharmacy & Science 1960. Owner and manager Grettenberger Pharmacies from 1932 to 1968. Operator of 450-acre farm and President of Lorann Flavoring Oil, Inc. Past President National Association Boards of Pharmacy. President Michigan Harness Horseman's Association. Past President Michigan Postmasters Association. Appointed to Mackinac Bridge Authority April, 1970.

MARY P. SHARP

Born in Ann Arbor. B.A. University of Michigan 1937, J.D. 1939, admitted to Supreme Court practice 1960. East Lansing City Council and Mayor Pro tem. Instructor, Business Law, Lansing Community College. President Lansing Community Services Council, Michigan Childrens Aid Society—Lansing. Member Michigan Fair Employment Practices Commission, East Lansing Human Relations Commission. Life member NAACP. Appointed to the Authority January, 1978.

BEN W. CALVIN

Born in Topeka, Kansas. Attended University of Illinois. Engaged in the cement business and came to Bay City in 1936 as manager of the Aetna Portland Cement Company. Shortly thereafter he was named President. Mr. Calvin has long been active in the Republican party and served ten years on the Republican State Central Committee. He was appointed to the Authority in 1964 and served until 1978.